Fluctuations and Scaling in Biology

Fluctuations and Scaling in Biology

edited by
Tamás Vicsek

OXFORD

UNIVERSITY PRESS

Great Clarendon Street, Oxford OX2 6DP

Oxford University Press is a department of the University of Oxford.
It furthers the University's objective of excellence in research, scholarship,
and education by publishing worldwide in

Oxford New York

Athens Auckland Bangkok Bogotá Buenos Aires Cape Town
Chennai Dar es Salaam Delhi Florence Hong Kong Istanbul Karachi
Kolkata Kuala Lumpur Madrid Melbourne Mexico City Mumbai Nairobi
Paris São Paulo Shanghai Singapore Taipei Tokyo Toronto Warsaw

with associated companies in Berlin Ibadan

Oxford is a registered trade mark of Oxford University Press
in the UK and in certain other countries

Published in the United States
by Oxford University Press Inc., New York

A catalogue record for this book is available from the British Library

Library of Congress Cataloging in Publication Data

Fluctuations and scaling in biology / edited by Tamas Vicsek.

Includes bibliographical references and index.
1. Biometry. 2. Fractals. 3. Biophysics.
4. Self-organizing systems. 5. I. Vicsek, Tamás.

QH323.5.F59 2001   570'.1'5195   2001016281

ISBN 0 19 850789 5 (Hbk)
     0 19 850790 9 (Pbk)

Typeset by Newgen Imaging Systems (P) Ltd., Chennai, India
Printed in Great Britain
on acid-free paper by
T.J. International Ltd., Padstow, Cornwall

Preface

Interdisciplinary research is becoming more widespread than ever. Most of the newly built research centres are intended to host investigations originally belonging to such diverse branches of sciences as biology, physics and chemistry. This is in part due to the fact that the techniques used in these branches of sciences have become sophisticated to a degree which makes the interaction of the groups using these methods necessary on an everyday basis.

One flourishing area is biological physics, where the collaboration between physicists applying the most recent theoretical and experimental techniques to biology and life scientists interested in making use of these approaches has led to a number of important new results about life.

In particular, during the last decade the well-established tools of *statistical physics* have been successfully applied to an increasing number of biological phenomena. Statistical physics is a fruitful approach to systems dominated by fluctuations and/or many identical (or nearly identical) units. There are many such systems in biology. For example, the number of individuals in a group of organisms or the codons in a genetic code can be very large, and the behaviour of the macromolecules (e.g., molecular motors) is largely influenced by the thermal fluctuations.

My intention was to provide a book summarizing the majority of the most recent approaches and concepts born in the studies of biological phenomena involving collective behaviour and random perturbation and, in addition, to present some of the most important new results for more specialized readers. Thus, it is hoped that the book will be able to serve as a text for studying scaling and fluctuations in biology, hence the emphasis is on presenting results in a reproducible manner rather than on briefly reviewing a large number of contributions. Obviously, the field is too broad to enable all of the related topics to be included. Among the important aspects not treated here are, for example, scaling as a function of the body size, heart rate variability and scaling of patches of vegetation.

Collaboration with many colleagues has greatly helped me in gaining an insight into the phenomena discussed in this book. I learned a great deal about biological systems from working together with A-L. Barabási, E. Ben-Jacob, I. Cohen, A. Czirók, M. Cserzö, I. Derényi, D. Helbing, E. Madarász, K. Schlett, O. Shochet, and H. E. Stanley and I thank them for this. I am also grateful to T. Biró, F. Family, V. Horváth and J. Kertész for many useful discussion and collaboration related closely or indirectly to the subject of this book.

My thanks to the co-authors/contributors of the chapters/sections in this book for producing a high-quality text on their part on time. I am grateful to Attila Csótó for his important help with the technical aspects of the editorial work.

Finally, this Preface provides me with a good opportunity to express my gratitude to my wife, Mária Strehó, who has helped me in many ways in producing this book.

Budapest, April 2000

Tamás Vicsek
Editor

Contents

List of Contributors

Z. Csahók

Department of Biological Physics, Eötvös University, Budapest, Pázmány P. Stny 1A, H-1117, Hungary.

A. Czirók

Department of Cell Biology and Anatomy, Kansas University, Medical Center, 1008 WHW 3901 Rainbow blvd., Kansas City, KS 66160-7400, USA. **aczirok@kumc.edu**

I. Derényi

Institute Curie, UMR 168, 26 rue d'Ulm, F-75248 Paris 05, France. **derenyi@angel.elte.hu**

D. Helbing

Institute for Economics and Traffic, Faculty of Traffic Sciences "Friedrich list", Dresden University of Technology, D-01062 Dresden, Germany. **helbing@trafficforum.de**

M. Molnár

Institute for Psychology, Hungarian Academy of Sciences, POB 398 Budapest, Hungary. **molnar@cogpsyphy.hu**

P. Molnár

Institute for Psychology, Hungarian Academy of Sciences, POB 398 Budapest, Hungary.

T. Vicsek

Department of Biological Physics, Eötvös University, Budapest, Pázmány P. Stny 1A, H-1117, Hungary. **vicsek@angel.elte.hu**

Introduction

T. Vicsek

Recent discoveries in biology have motivated an increasing number of physicists to study biologically inspired problems. It has turned out that the methods developed by physicists can be very useful in making the investigations in life sciences more quantitative. In turn, there is a growing interest in the biology community to present and interpret their findings by using the advanced hardware and the numerical and theoretical techniques developed by physicists.

In particular, during the last decade the well-established tools of statistical physics have been successfully applied to an increasing number of biological phenomena. This has been in part due to a breakthrough in statistical physics. In simple terms, statistical physicists have been able to theoretically establish that physical systems consisting of a large number of interacting particles obey universal laws that are independent of the microscopic details. The concepts are so general that they can be applied to a broad class of systems even beyond physics. Examples include the dynamics of companies and finances, genetic coding, cooperativity in granular media, ageing, bioevolution, polymers, traffic flow (e.g., see [1–3]).

Statistical physics is a useful approach to the investigation of a system if it has many identical (or nearly identical) units and/or the random perturbations play an important role. There are many such systems in biology. For example, the number of individuals in a group of organisms or the codons in a genetic code can be very large, and the behaviour of the macromolecules (e.g., molecular motors) is largely influenced by the thermal fluctuations.

The microscopic laws are known in physics, while for systems in biology even the laws of interaction are far from being exactly revealed. However, cooperativity of the many interacting units introduces in all cases new qualitative features. The various characteristics of the associated collective behaviour include transitions between disordered and ordered structures, large fluctuations in space and time, and long-range correlations. Critical behaviour was originally a purely equilibrium notion, however, nonequilibrium systems are in many cases also characterized by a critical-like scale-invariant regime, at least between some natural cutoffs.

There are important analogies and differences between physical and biological systems. As an example, let us consider motion. In physics the origin of motion is usually some external force. In mechanics the basic task is to identify these forces and the laws related to them. One always considers closed systems for Newton's second law to be applicable. In the world of physics this method is perfectly appropriate. On the other hand, when it comes to living systems it may not be possible to include all the factors which make the system closed. To illustrate the situation consider the problem of flocking birds. This phenomenon can be understood, with some abstraction, without the detailed description of what forces are exerted by a bird and exactly which metabolic pathways are involved in generating forces. Of course, more detailed models possess more predictive power. In biology, however, including into a model the many important details of a phenomenon makes the model practically untreatable. This is due to the enormous complexity involved in biological processes. People probably thought in the medieval ages the same of celestial problems. Today we can easily handle celestial mechanics, but we can only hope that there exists a common theoretical framework for living systems.

On the other hand, to a first approximation a system of organisms can be considered as an open interacting multi-particle physical system. Then, one can attempt to apply the methods recently developed in the investigations of complex systems.

In the case of collective motion, for example, the concept of self-propelled systems has been proposed. Such systems consist of 'particles' which are capable of moving due to internal processes. Formally such systems violate the conservation of momentum, but one should always keep in mind that these systems, in contrast to usual physical systems, are open and far from equilibrium. Interacting self-driven systems exhibit a range of phenomena: intracellular motion, motion of bacteria, schooling of fish, flocking of birds, swarming of insects, humans rushing out of a stadium after a soccer match, or cars on a highway. Methods based on self-driven individuals (or agents as they are called) are also extensively used, both in artificial-life research and in computer graphics.

In this book we review a selection of biological phenomena which can be interpreted using methods based on techniques developed in statistical physics. In Chapter 1 a brief overview of the general features of complex patterns, fluctuations and scaling are given. In Chapter 3 the concept and some applications of self-organized criticality are presented. Chapter 4 contains a number of contributions related to the complex spatial and temporal patterns emerging in biological systems containing many units. The following chapter is devoted to the theoretical studies of the microscopic mechanisms of biological motion. This is a newly emerging field where the effects of the microscopic fluctuations in the transport processes within a cell is being clarified. Finally, in Chapter 6 we discuss the main aspects of the collective motion of organisms including the patterns developed by humans as they move together.

References

[1] Iannaconne, P. M. and Khokha, M. (ed.). *Fractal Geometry in Biological Systems*. CRC Press, New York, 1996.

[2] Hu, C-K. and Leung, K-T., (ed.). *New Directions in Statistical Physics*. Proceedings of the IVth Tapei Int. Symposium. North-Holland, Amsterdam, 1998.

[3] Gervois, A., Iagolnitzer, D., Moreau, M. and Pomeau, Y. (ed.). *Statistical Physics: Invited papers from STATPHYS20*. North-Holland, Amsterdam, 1999.

1 Basic concepts

T. Vicsek

The complexity of biological systems is manifested in many ways. Here we shall consider those aspects of life which involve random fluctuations and a hierarchical underlying structure resulting in a power-law dependence of the various quantities characterising these systems. As will be shown, these two features – fluctuations and scaling – are frequently and intimately related, although in some cases they appear independently of each other.

Are fluctuations an important, inherent ingredient of life? What is their origin and impact? These are important questions which have to be addressed before we lead the reader to the more advanced parts of the text.

It is almost a trivial statement that nearly all processes in biology involve randomness beyond a negligible level. On the other hand, except for some artificial situations, all phenomena in nature have the elements of stochasticity in them. There is no such a thing as a noiseless or absolutely deterministic system and this is true from the behaviour of galaxies down to elementary particles. In the intermediate size range the presence of temperature fluctuations, the various nonlinearities/instabilities represent the main source of randomness. This is known to be so and is understood in many cases for non-living systems.

Biological systems are not exceptions from the same rule. Perhaps the most typical feature of a biological object or signal is that they simultaneously possess some specific pattern and deviations around these patterns. If we take as an example an animal, let us say a dog, we can consider the following illustrations of the above statement: (i) no two dogs are exactly the same, but they are similar in many ways; (ii) they have typical reactions, but never react in a perfectly identical way; (iii) their heart beats almost regularly, but a closer analysis shows specific fluctuations around the average heart beat rate; (iv) if we study the electric signals (with the help of an electrode) of their brain, we see an almost randomly looking series of spikes; (v) if we happened to see the motion of the individual cells or even large molecules in their body (e.g., blood cells, sperms, RNA molecules), we would observe an erratic behaviour around an average tendency. The list of such examples could be continued for long.

As will be discussed later, in many cases the above-mentioned stochastic changes or fluctuations are not completely random and can be associated with power-laws or scaling.

In short, if a system is made of many interacting units, specific statistical features involving fluctuations and scaling emerge.

Some important aspects of fluctuations when the origin of an apparently random behaviour is relatively well understood are listed below.

1. Microscopic objects are subject to so-called thermal fluctuations. It is a fundamental feature of all systems that if they have a well-defined temperature T, than each microscopic particle (atom or molecule) in them has in average an amount of kinetic energy $\frac{1}{2}kT$ (k is the Boltzmann constant) for each degree of freedom (mode of motion). A more complex molecule has many degrees of freedom, and is also subject to 'kicks' from the neighbouring molecules moving with a velocity corresponding to their kinetic energy. As a result, an individual molecule interacting with many others follows a more or less random trajectory due to the many subsequent, randomly occurring 'kicks' from their neighbours. It is a beautiful subject to understand how such fluctuations are tamed by the specific processes (diffusion, enzymatic interactions) inside living cells and result in an organized (with some fluctuations) behaviour.

2. When many similar, but not necessarily microscopic, objects (biological or non-living) are present in a system, there are further reasons to consider the random aspect of the behaviour.

    In general, if many moving objects interact, the motion of the individual objects is bound to be random-like (as in the above-mentioned microscopic case), even if the motion or the interaction is deterministic and we consider macroscopic objects. The trajectories are subject to many changes of varying degree. Instead of looking at such processes as deterministic it is conceptually more useful to assume that the motion is stochastic, and it is the statistical features of the assembly of objects which should be determined when describing the system. It is not only the direction of motion which can be changed during interaction. For objects with a directedness their direction can also be modified due to some direction-dependent interaction.

    Now, as a result of these interactions, various spontaneous processes may take place in the system: for attractive forces, aggregates are likely to form, or groups with the same directedness of their members may appear. A school of fish is a common example of aggregation. However, due to the complex nature of a system with many objects moving both randomly and with some order (in a group, but with some perturbations) another kind of fluctuation, a random distribution of group sizes, is produced. Often the distribution of group sizes follows a *power-law*, or in other words, a *scaling distribution*.

3. Nonlinearities are known to lead to a very complex behaviour which – especially in the presence of thermal fluctuations – can be considered as random. A common spatial example is the formation of branching patterns under conditions leading to the unstable growth of more advanced branches. In these cases the smallest

perturbations are likely to result in a new, quickly growing side-branch and the structure attains the well-known branching morphology so common in biology (e.g., trees or blood vessel networks). Although such networks possess some specific features typical for the given biological object, they are also irregular. In addition, as will be shown, they have a specific hierarchical structure best described by fractal geometry.

In the following parts of this introductory chapter I briefly discuss the basic concepts related to fluctuations and scaling in biology and summarize the most important findings obtained in the related investigations.

1.1 Fluctuations

1.1.1 Noise versus fluctuations

The origin of fluctuations can widely vary. In most cases, however, they are due to the above-mentioned 'thermal noise' or erratic motion of microscopic particles. Typically, the noise is not correlated, which means that the value of the fluctuating quantity F at the location r at time t does not depend on its value at a different location and at an earlier moment of time. Symbolically we write the correlation function in the form of a delta function

$$c(r) = \langle F(r,t)F(r',t') \rangle - \langle F(r,t) \rangle \langle F(r,t) \rangle = C\delta(r - r', t - t')$$

where the δ function is equal to zero for any non-zero values of its arguments and C is some constant. The averaging (denoted by $\langle \ldots \rangle$ is made over all values of the arguments. This expression holds for the uncorrelated white or shot noise, while for correlated fluctuations $c(r)$ has more complex forms.

Fluctuations can be more complex than just white noise. Many times they represent an inherent, characteristic feature (reaction to white noise) of the system itself. For example, due to the thermal (white) noise the fluctuations in the magnetization of a ferromagnet near its critical point can strongly increase and exhibit specific correlations.

Noise and fluctuations play a central role in ordering phenomena. A system of many interacting units with an interaction 'trying' to force the units to behave in the same way in the presence of strong external fluctuations (noise) may not be able to order. On the other hand, as the magnitude of noise decreases (e.g., the temperature is lowered), the objects in the system may already assume their common or collective pattern of behaviour, for example, they spontaneously develop a common direction of motion or find their right place for a crystalline structure. These noise-driven transitions will be discussed briefly in the scaling part of this introductory chapter.

In the following, when only their stochastic nature is relevant, we use noise, fluctuations and random perturbations as synonyms. However, random changes appearing as an inherent behaviour (response) of the system itself will always be called fluctuations.

A further aspect of fluctuations involves transport processes in the presence of noise. Interestingly, even uncorrelated fluctuations (these are random changes without any tendencies) may result in a behaviour with a well-defined tendency. This happens in the case of *molecular motors*, where white noise assists the motor proteins to proceed along specific intracellular tracks. On the other hand, white noise alone cannot produce currents, as this would contradict the second law of thermodynamics. Here we mention the basic findings about this fascinating new direction in biological physics.

1.1.2 Molecular motors driven by noise and fluctuations

In the inorganic world transport always takes place along a macroscopic gradient of a potential (or an analogous quantity). Things fall down due to the gravitational force which can be obtained from the derivative of the gravitational potential. Even the global transport of microscopic objects such as molecules takes place along the extended gradient of the so-called chemical potential. For example, particles tend to diffuse from a denser region to a less dense one, covering a macroscopic overall distance if the gradient of the density extends over that distance. Electrons move in a wire following the gradient of the electric potential, which is larger at one end of the wire than at the other end.

This is not how transport is achieved in most biological systems. The above mechanism tends to bring a system into a motionless state: as the objects move along the potential gradient they simultaneously decrease the value of the overall gradient. For example, the difference in the concentration (driving diffusional transport) decreases with time as the particles diffuse to the spots of smaller concentrations (and increase the density at these spots). Instead, life is about generating differences, building structures from a less-patterned environment.

One mechanism for doing this is *motion along periodic, but locally asymmetric structures*. Imagine a sawtooth-like (or ratchet) potential: it is piecewise linear, with two different gradients (slopes). We also have a particle in this potential, most of the time 'sitting' in one of the minima ('valleys'). Now, if we pull this particle periodically back and forth in this sawtooth-like potential, the following cases are possible: (i) the force is large enough for the particle to be pulled out from a minimum both to left and right, (ii) the force is strong enough only to pull out the particle in the direction of the smaller gradient (steepness), or (iii) the particle is not pulled out from the valley because the force is too weak.

Obviously, case (ii) establishes a situation in which transport is possible without any global potential difference: the particle moves in the direction of the smaller slope although the force acting on it is zero on average (acts back and forth). In this way particles can be collected at one end of a track with such a potential, and thus a concentration difference, i.e., a structure, can be built up.

However, life is not so simple. Such processes occur at the molecular level, where *fluctuations are very strong* for two reasons: (a) the particle (called motor protein) is kicked by the other molecules in the system randomly, in a noisy, uncorrelated manner all the time, (b) the periodic, deterministic back-and-forth driving cannot be established in a microscopic environment as well – instead, the energy supplied by the so called ATP molecules and providing the conformational changes of the molecular motor resulting in its tendency to move back and forth also arrives stochastically (the ATP molecules are 'consumed' at times which follow a Poisson distribution).

The picture which emerges is the following: a possible, simplified representation of biological motion is that of the motion of a Brownian particle in an asymmetric periodic potential. The corresponding equation has to account for the stochastic nature of the process: both for the white noise coming from the environment and for the irregular nature of the energy input. *The related Langevin equation approach is discussed in detail in Chapter 5.*

There is one interesting point here: in the case of motion along asymmetric periodic potentials noise may play a role in *enhancing the transport*. This is rather paradoxical, we are more used to the notion that random perturbations typically destroy tendencies. In the case of molecular motors, however, it may happen that adding white noise results in a stronger current. The easiest way to show this is as follows. Imagine case (iii) when the external force pulling the particle back and forth is not strong enough for the particle to escape in any of the two directions. Now, if we add noise (a randomly changing small amount of extra force), in some cases the overall force may exceed the critical value necessary to pull out the particle from a valley. This will happen more frequently in the direction of the smaller slope, since the critical amount of force is smaller in that direction, and there will be an overall current in the direction of the smaller slope. This is why such systems are also called *thermal ratchets*.

Another interesting variant of this situation is when we consider one single force (instead of the sum of a deterministic periodic back-and-forth acting force and an uncorrelated white-noise one) changing stochastically. Now, if this single fluctuating force is completely uncorrelated, or in other words, thermal or white noise, then no global transport is possible. If the fluctuations are thermal noise-like, than the system is in equilibrium and no transport is possible in thermal equilibrium.

On the other hand, if the fluctuating force (noise) *is correlated*, transport already becomes possible. The probability to leave in the less steep direction will be still larger than in the opposite direction. In this way, our ratchet 'rectifies' the fluctuations: *it is able to make use of its non-white part.*

1.2 Scaling

A quantity F *scales* as a function of its argument x if changing the argument by a factor (e.g., changing x to Ax) does not change the form of the functional dependence of F on x (apart from a constant factor). This is trivially so for a function of the form of a power-law, but is not, as a rule, true for other functions. For example, $F = x^2$ scales because $F' = F(Ax) = A^2x^2 = A^2F(x)$, while $F = \log(x + B)$ does not scale because in this case $F' = F(Ax + B)$ cannot be reduced to a form containing $F(x)$ and a constant factor only.

The scaling quantities we shall consider are mostly of stochastic nature. Thus, the specific functional dependences will be valid for the average of the given quantities as a function of their arguments. Each realization of some stochastic process has a fluctuating outcome, but making an average over several processes, or over a single process having sequences fluctuating around an average, can provide the proper estimate of the quantity of interest.

1.2.1 Critical behaviour

Perhaps the most typical collective phenomenon exhibited by an assembly of many interacting particles is the so-called *phase transition*, when, as a function of an external parameter (like temperature), the particles collectively change their behaviour. For example, during freezing, all of the molecules of a fluid move to a specific position, so that the resulting structure becomes a crystal with regular microscopic structure.

In the vicinity of such transitions interesting spatial and temporal correlations can be observed in the systems and these features will be relevant for the majority of the topics discussed in this book.

In particular, during second-order phase transitions the so-called 'critical state' (or critical phenomena) can be observed in which the ordinarily exponential functional dependences are replaced by an algebraic (power-law) dependence of the relevant quantities on their parameters. A *power-law dependence* of the quantity $n(s)$ (e.g., the number of schools containing s fish) is of the following form: $n(s) \sim s^{-\tau}$, where $\sim$ expresses proportionality, and τ is some exponent. The power-law dependence is very special: for example, a power-law decay of the number of clusters (schools of fish) as a function of their size (number of fish in a school) means that very large clusters may occur with a probability which is not negligible (this probability would be extremely small if the number of clusters decreased exponentially with growing cluster size, as it does for regular states). If a quantity changes according to a power-law when the parameter it depends on is growing linearly, we say it *scales*, and the corresponding exponent is called a *critical exponent*.

Why are such states called critical? Because they are extremely sensitive to small changes or perturbations. If a human being is in a critical state, it means his or her state can get worse very easily. In the case of lattice models a small change in the temperature may lead to the quick collapse or birth of very large clusters. In the last two decades statistical physicists have worked out delicate theories and methods to interpret the behaviour of such critical transitions and states, and in the following sections we shall consider the application of the related concepts to biological phenomena involving many similar units.

The important point is that scaling typically involves *universality*: instead of particles we can imagine similar organisms. If the interaction among these organisms is relatively simple, and is analogous to those which produce scaling or phase transitions in non-living systems, than we can expect the same type of behaviour in such systems of living entities as well.

1.2.2 Scaling of event sizes: Avalanches

As mentioned above, scaling can be observed during an equilibrium phase transition, but in the following we shall argue that a power-law dependence of the various important quantities can emerge in the non-equilibrium world (of life) as well. In fact, it is the *non-equilibrium state* in which structures can emerge spontaneously from an originally homogeneous medium.

A particular and important departure from equilibrium is when the system is 'slowly driven' to a stationary state. Slow driving may mean the gradual addition of some quantity (energy) to a system which may also loose this energy due to interactions. If the interaction between two parts of the system is such that a change exceeding a critical value of the given quantity in one unit results in a similar exceeding of the same critical value in the neighbouring unit, than *large, avalanche-like series of changes* may take place in the system when it is close to a (critical, balanced) state. As this state is both spontaneously emerging and critical, the associated phenomenon is called self-organized criticality (SOC). In this state the system is very sensitive to fluctuations, since a small perturbation may lead to a large avalanche. In this sense the system is in a critical state. Not all of the avalanches are large, the majority of them are small, but the probability of having a large avalanche does not go to zero very quickly with the avalanche size. In many such slowly driven systems scaling (power-law) of the avalanche size distribution can be observed. Avalanches are sometimes very large, as we know from the news on skiing areas (in this context an earthquake is also an avalanche), and they are the so-called big events in the theory of slowly driven systems.

The simplest example is that of a *sandpile*. Imagine that we add grains of sand to a growing pile. As the grains are dropped, most of the time the surface of the pile becomes only slightly rearranged. However, from time to time the new grain triggers a long series of events: grains rolling down the slope drag many more grains with them. A *simple model* along a line segment would contain columns of particles. A new particle is dropped at a random position. If the height difference between two columns becomes larger than two particles, then from the higher column two particles are removed and added to the two neighbouring column. Particles at the edge of the segment drop out from the system completely.

In this book two examples for biological SOC are discussed. The structure of the lung is such that it can be brought into analogy with the sandpile model. The air entering the lung has to go through a sequence of airways, each opening if the pressure exceeds a critical value. By forcing the air to gradually enter the experimentally investigated lung, large jumps in the terminal airway resistance have been observed. These jumps corresponded to avalanches: the subsequent opening of a large set of airways in a short time. The distribution of jumps followed a power-law. The observation of strongly fluctuating extinction rates and the corresponding SOC-related theory is also discussed in Chapter 3.

1.2.3 Scaling of patterns and sequences: Fractals

Nature is full of beautiful complex shapes which are far more intricate than the ideal-ized forms proposed by Euclid more than two thousand years ago. This is particularly true for the living world, where complicated structures are generated during embrio-genesis. Many of these patterns are random branching networks; examples include trees, the network of blood vessels, airways in the lung, neural nets, etc. These highly hierarchical patterns can be best interpreted in terms of fractal geometry.

Fractals are fascinating geometrical objects characterized by a non-trivial fractional dimension. Imagine a growing pattern whose mass M (the number of particles it contains) increases more slowly than the d-th power of its radius R, where d is the dimension of the space in which the pattern is developing. This is clearly different from the case of homogeneous structures that we are used to. For fractals $M(R) \sim R^D$, where D is called the fractal dimension since in many cases it is *not an integer*, but a fractional number less than d. If the above relation is true for a pattern, it is bound to be *self-similar* in the sense that a small part of it looks the same as the whole structure after it is expanded isotropically. For deterministic mathematical fractals the blown-up piece looks exactly the same as the whole object. For random patterns self-similarity is satisfied in a stochastic manner. The fractal dimension can also be defined through the expression $c(r) = 1/N \sum_{r'} \rho(r)\rho(r + r') \sim r^{D-d}$, where $c(r)$ describes the density–density correlations within the pattern and $\rho(r)$ is equal to unity if there is a particle at the position r and is equal to zero otherwise. For isotropic structures the correlation function $c(r)$ is equivalent to the probability that one finds a particle belonging to the cluster at a distance $r = |r - r'|$ from a fixed point on the cluster. In this case an averaging can be made over the directions as well.

The meaning of the above statements is that fractals can be looked at as structures exhibiting *scaling in space* since their mass as a function of size or their density as a function of distance behave as a power-law.

Self-affine structures represent another type of fractal. For such objects a small part of the fractal must be enlarged in an anisotropic way to match the entire pattern. For example, if the fractal is embedded into two dimensions, for self-affine fractals one achieves matching by rescaling the size horizontally and vertically by different factors.

Fractal Bacterial Colonies

Perhaps the best-defined biological systems exhibiting fractal growth are bacterial colonies. By careful control of the experimental conditions it has been possible to obtain very reproducible results on the development of complex branching patterns made of many millions of bacteria as they multiply on the surface of an agar (gel) layer in a Petri dish. The related beautiful patterns are discussed in Chapter 4.

Typically, bacterial colonies are grown on substrates with a high nutrient level and intermediate agar concentration. Under such 'friendly' conditions, the colonies develop simple (almost structureless) compact patterns with a relatively smooth envelope. This behaviour fits well the contemporary view of bacterial colonies as a collection of independent unicellular organisms. However, in nature, bacterial colonies regularly must cope with hostile environmental conditions. What happens if we create hostile conditions in a Petri dish by using, for example, a very low level of nutrients or a hard surface (high concentration of agar), or both? The bacterial reproduction rate, which determines the growth rate of the colony, is limited by the level of nutrient concentration available for the bacteria. The latter is limited by the diffusion of nutrients towards the colony. Hence, the growth of the colony resembles *diffusion-limited growth* in inorganic systems leading to fractal patterns.

Diffusion-limited growth leads to random branching patterns because of the following *instability*: if a given part of the growing surface is slightly more advanced than the surrounding region, this part will advance faster than the neighbouring parts of the colony, because it will be closer to the source of the nutrient diffusing from the outer regions of the Petri dish. In turn, parts lagging behind tend not to grow any more, since in those regions no nutrient will be available as the nutrient diffusing towards the colony will be consumed by the advanced parts of the colony. This is positive (negative) feedback: a protrusion grows faster (and produces a branch) the screened fjords stop to grow completely. The resulting pattern has a radially growing treelike structure.

In reality the situation is somewhat more complex, since the bacteria can communicate through *chemotaxis*. They are able to pass on information about their environment and increase/decrease the growth rate at other points in the colony. The communication enables each bacterium to be both actor and spectator (using Bohr's expressions) during the complex patterning. The bacteria developed a particle–field duality: each of the bacterium is a localized (moving) particle which can produce a chemical and physical field around itself. For researchers in the pattern formation field, the above communication regulation and control mechanism opens a new class of tantalising complex models exhibiting a much richer spectrum of patterns than the models for inorganic systems.

All this can be investigated by constructing suitable *computer models*. In the corresponding calculations a number of factors are taken into account (see Chapter 4); good agreement with the experimental observations can be achieved by assuming a nutrient-dependent multiplication rate, diffusional motion of the bacteria on the agar surface, chemotactic signalling, etc. These simulations are different from the common approaches in physics and biology. Physicists prefer to build simple models ignoring many of the details and look for universal behaviour. Biologists mostly use specific models reflecting the biological details of the system under investigation. The models used to mimic bacterial colony development in the computer interpolate between these approaches and are *aimed at finding universal behaviour taking into account most of the biologically relevant details*.

Correlations in the genetic code

One possible representation of the vast information stored in the extremely long sequences of DNA data is a random walk built to correspond to such sequences. In this approach DNA is mapped onto a process which can be regarded as a walk: each of the four 'letters' of a DNA sequence is identified with a step in a given direction. Then, the specific features of this walk can be analysed using methods borrowed from statistical physics.

Given the walk, one can look for correlations. Two series of data (X and Y) are correlated if there is a relationship between the corresponding elements of the series. When searching for correlations within a single sequence of data we can ask how the value X_i is related to the value X_{i+j}. By comparing the two values with the average of X one can get information about the question whether two values in the data set separated by j elements are correlated.

An ordinary random walk has no long-range correlations. One of the most relevant questions one can raise in the context of DNA sequences is the location of coding and non-coding parts in the genome. In the case where these two kinds of subsequences have different kinds of correlations we may be able to differentiate between the coding and non-coding part without any prior knowledge about the sequences. Indeed, it has been shown that the random walks corresponding to *non-coding parts have long-range correlations* in contrast to the coding parts (which have short-range correlations only).

As an alternative to the DNA walk, the symbol sequence corresponding to a DNA molecule can be regarded as a written text composed by using four letters. Since we do not know the 'language' of the text, we have to apply methods developed for analysing written (natural) texts of unknown origin. In particular, one can ask the question whether two texts were written in the same language or not. We expect larger correlations between text of the same origin (language). Here two sequences are correlated if the scalar product of the two appropriately defined vectors (corresponding to them) has a value different from that it would have for two uncorrelated sequences. To a first approximation this method is language insensitive.

When applying the vector space technique to DNA sequences, in a way we look at DNA as an encoded text written in an unknown language. Nevertheless, we expect to locate correlations between parts of the sequences due to similarities in their underlying structure. In the case where the language of the coding parts is different from the non-coding ones, we get a higher value for the corresponding scalar product (as has been demonstrated in some cases).

1.2.4 Scaling in group motion: Flocks

Group motion (flocking) is a beautiful phenomenon many times capturing our eyes. Here flocking is understood in the general sense of the word, including herding of quadrupeds, schooling of fish, etc. In the last chapter of this book we address the question whether there are some global and universal, transitions in flocking when many organisms are involved and such parameters as the level of perturbations or the mean distance of the organisms are changed.

Everyone has experienced how an initially randomly directed group of birds feeding on the ground *is spontaneously ordered into a well-organised flock* when they leave because of some external perturbation. This ordering is a highly non-trivial question since in a huge flock of several hundred or thousand birds there is usually no 'leader' bird (we do not consider here the V-shaped or other structured flight of some large-bodied birds) and not even all birds can visually interact. Still, the whole flock selects a well defined direction. Such ordering is familiar in the equilibrium phase transitions of magnetic systems, and the corresponding findings may provide clues to the understanding of the more complex far-from-equilibrium ordering of moving organisms.

Self-propulsion is an essential feature of most living systems. In addition, the motion of the organisms is usually controlled by interactions with other organisms in their neighbourhood and randomness plays an important role as well. It is possible to design simple computer models which simulate the collective motion and take into account the most relevant ingredients of the phenomenon.

A *simple model of collective motion* consists of particles moving in one, two or three dimensions. The particles are characterized by their (off-lattice) location $\mathbf{x}_i$ and velocity $\mathbf{v}_i$ pointing in the direction ϑ_i. To account for the self-propelled nature of the particles the magnitude of the velocity is fixed to v_0. A simple local interaction is defined in the model: at each time-step a given particle assumes the average direction of motion of the particles in its local neighbourhood with some uncertainty. Such a model is a transport-related, non-equilibrium analogue of the ferromagnetic models. The analogy is as follows: the function tending to align the spins in the same direction in the case of equilibrium ferromagnets is replaced by the rule of aligning the direction of motion of particles, and the amplitude of the random perturbations can be considered proportional to the temperature.

In addition, collective motion can be described by continuum equations. The collection of 'birds' is then looked at as particles in a fluid subject to fluctuations and satisfying the condition of trying to move with a given velocity.

Both theoretical approaches led to the conclusion that there are interesting, in some cases unexpected (compared to equilibrium systems), transitions in collective motion. For example, if the noise (level of perturbations, corresponding to temperature in the case of ferromagnets) is decreased, the *originally disordered flock becomes ordered in analogy with second-order phase transitions*. The level of global order, the fluctuations around this order and several related quantities all scale, i.e., behave according to a power-law as a function of the distance from a critical level of perturbations.

Pedestrian simulations

A special kind of flock is a group of people. In the last chapter interesting applications of pedestrian simulations are discussed. Just as in the case of other organisms, people can be represented by particles 'dressed' by the appropriate interactions. Simulations of humans moving in confined places lead to a number of interesting effects reproducing related observations.

Freezing by heating

Freezing by heating is an effect observed when the particles are driven in opposite directions. The related simulations demonstrated that more nervous or hectic changes (heating) of the direction of motion can cause a breakdown of an efficient pattern of cooperative interactions and finally produce a deadlock (freezing). In particular, this may be relevant for panicking pedestrians in a smoky environment, who tend to build up fatal blockings. The system described in the last chapter consists of a mesoscopic number of driven particles with repulsive hard-core interactions moving into opposite directions under the influence of fluctuations. An example of such a system is pedestrians walking in a passage.

In short, 'freezing by heating' means a transition from a fluid state (with self-organised lanes of uniform direction of motion) to a solid, crystallized ('frozen') state just by increasing the noise amplitude ('temperature'). This is in contrast to, for example, melting, where increasing the temperature increases the energy and order is destroyed, and to noise-induced ordering in glasses or granular media, where increasing the temperature drives the system from a disordered metastable state (corresponding to a local energy minimum) to an ordered stable state (corresponding to the global energy minimum). Instead, 'freezing by heating' shows an increase in the order at increasing temperature, although the total energy increases at the same time. The crystallized state can also be destroyed by *ongoing* fluctuations with extreme noise amplitudes giving rise to a third, disordered ('gaseous') state with randomly distributed particles. Thus, with increasing 'temperature' θ, we have the atypical sequence of transitions *fluid* $\longrightarrow$ *solid* $\longrightarrow$ *gaseous*.

Further variants of pedestrian simulations allow *the quantitative investigation of trail formation, optimization of passage geometries,* etc.

In this chapter, I have attempted to present in a simplified manner a *selection of concepts, topics and results discussed in much more detail in the main body of the book*. For details necessary for a deeper understanding of the concepts and findings related to fluctuations and scaling in biology I advise the reader to read other parts of our book as well (where the related references are also given).

2 Introduction to complex patterns, fluctuations and scaling

In this book we mostly consider models of reality, since 'reality' in the case of biology is far too complex to allow a complete theoretical treatment. On the other hand, whenever it is possible, we are trying to present models as realistic as possible in order to reflect the essential features of the specific phenomena occurring in nature. In many cases we are concerned with *systems consisting of many similar objects* and this feature has particular implications for the kinds of models we consider.

Various models allowing exact or numerical treatment have been playing an important role in the studies of biological processes. Because of the complexity of the phenomena it is usually a difficult task to decide which of the many factors influencing the processes are the most significant. In a real system the number of all possible factors can be too large; this number is decreased to a few by appropriate model systems. Thus, the investigation of these models provides a possibile way to detect the most relevant factors and demonstrate their effects in the absence of any disturbance.

Systems consisting of many similar units can be successfully described in terms of *particles*, where the word particle can stand for a molecule as well as for more complex objects, including organisms. Then, the particular nature of the model is given by the features of the individual particles and by the ways these particles interact with each other.

The spatial arrangement of these particles is frequently of major interest. Structures consisting of connected particles are usually called *clusters* or *aggregates*. In most cases the particles are assumed to 'exist' on a *lattice* for computational convenience, and two particles are regarded as connected if they occupy nearest neighbour sites of the lattice. However, for studying universality and related questions, *off-lattice* or further neighbour versions of clustering processes can also be investigated. A lattice site with a particle assigned to it is called occupied or filled. An important additional feature included in the majority of models to be described is stochasticity, which is typical for biological phenomena.

In this chapter we discuss the basic features of the complex patterns produced by a wide variety of biological growth processes. In many cases biological growth leads to random fractal structures characterized by a non-integer dimension defined below. In Section 2.2 the principles behind the motion in the presence of fluctuations will be presented. Since biological motion is produced by motor molecules acting on a microscopic scale, thermal noise and other stochastic perturbations are essential.

Finally, we discuss continuous phase transitions where the concept of scaling plays a central role. Scaling means a power-law dependence of a quantity on its argument and, as will be demonstrated, is a feature showing up in an unexpectedly large selection of biological systems. We know from the early studies of scaling in physics that it is a fundamental characteristic of a system. The power-law dependence of a quantity usually involves a similar behaviour of the other quantities in a system; in addition, the exponent (corresponding to the power law) is typically not sensitive to the details of the processes considered.

2.1 Fractal geometry

T. Vicsek

During the last decade it has widely been recognized by researchers working in diverse areas of science that many of the structures commonly observed possess a rather special kind of geometrical complexity. This awareness is largely due to the activity of Benoit Mandelbrot [1], who called attention to the particular geometrical properties of such objects as the shores of continents, the branches of trees, or the surfaces of clouds. He coined the name *fractal* for these complex shapes to express that they can be characterized by a *non-integer* (fractal) *dimensionality*. With the development of research in this direction the list of examples of fractals has become very long, and includes structures from microscopic aggregates to the clusters of galaxies. Objects of biological origin are many times fractal-like.

Before starting a more detailed description of fractal geometry let us first consider a simple example. Figure 2.1 shows a cluster of particles which can be used for demonstrating the main features of fractals. This object was proposed to describe diffusion-limited growth [2] and has a loopless branching structure reminiscent of many shapes of biological origin. Imagine concentric circles of radii R centered at the middle of the cluster. For such an object it can be shown that the number of particles in a circle of radius R scales as

$$N(R) \sim R^D, \tag{2.1}$$

where $D < d$ is a non-integer number called the *fractal dimension*. Naturally, for a real object the above scaling holds only for length scales between a lower and an upper cutoff. Obviously, for a regular object embedded into a d-dimensional Euclidean space Equation (2.1) would have the form $N(R) \sim R^d$ expressing the fact that the volume of a d-dimensional object grows with its linear size R as R^d. Clusters having a non-trivial D are typically *self-similar. This property means that a larger part of the cluster after being reduced 'looks the same' as a smaller part of the cluster before reduction.* This remarkable feature of fractals can be visually examined in Fig. 2.1, where parts of different sizes (included into rectangular boxes) can be compared from this point of view.

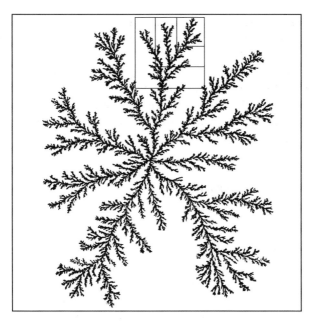

Fig. 2.1 A typical stochastic fractal generated in a computer using the diffusion-limited aggregation model.

2.1.1 Fractals as mathematical and biological objects

In addition to self-similarity mentioned above, a characteristic property of fractals is related to their volume with respect to their linear size. To demonstrate this we first need to introduce a few notions. We call *embedding dimension* the Euclidean dimension d of the space the fractal can be embedded in. Furthermore, d has to be the smallest such dimension. Obviously, the volume of a fractal (or any object), $V(l)$, can be measured by covering it with d-dimensional balls of radius l. Then the expression

$$V(l) = N(l)l^d \tag{2.2}$$

gives an estimate of the volume, where $N(l)$ is the *number of balls needed to cover the object completely* and l is much smaller than the linear size L of the whole structure. The structure is considered to be covered if the region occupied by the balls includes it entirely. The phrase 'number of balls needed to cover' corresponds to the requirement that $N(l)$ should be the smallest number of balls with which the covering can be achieved. For ordinary objects $V(l)$ quickly attains a constant value, while for fractals typically $V(l) \to 0$ as $l \to 0$. On the other hand, the surface of fractals may be anomalously large with respect to L.

There is an alternative way to determine $N(l)$ which is equivalent to the definition given above. Consider a d-dimensional hypercubic lattice of lattice spacing l which occupies the same region of space where the object is located. Then the number of boxes (mesh units) of volume l^d which overlap with the structure can be used as a definition for $N(l)$ as well. This approach is called *box counting*.

Returning to the cluster shown in Fig. 2.1 we can say that it can be embedded into a plane ($d = 2$). Measuring the total length of its branches (corresponding to the surface in a two-dimensional space) we would find that it tends to grow almost indefinitely with the decreasing length l of the measuring sticks. At the same time, the measured 'area' of the cluster (volume in $d = 2$) goes to zero if we determine it by using discs of decreasing radius. The reason for this is rooted in the extremely complicated, self-similar character of the cluster. Therefore, such a collection of branches is definitely much 'longer' than a line but has an infinitely small area: it is neither a one- nor a two-dimensional object.

Thus, the volume of a finite geometrical structure measured according to eqn (2.2) may go to zero with the decreasing size of the covering balls while, simultaneously, its measured surface may diverge following a *power-law* instead of the better behaving exponential convergence. In general, *we call a physical object fractal, if measuring its volume, surface or length with hyperballs of d, $d-1$, etc. dimensions it is not possible to obtain a well-converging finite measure for these quantities when changing l over several orders of magnitude.*

It is possible to construct mathematical objects which satisfy the criterion of self-similarity exactly, and their measured volume depends on l even if l or (l/L) becomes smaller than any finite value. Figure 2.2 gives examples of how one can construct such fractals using an iteration procedure. Usually one starts with a simple initial configuration of units (Fig. 2.2(a)) or with a geometrical object (Fig. 2.2(b)). Then, in the growing case this simple seed configuration (Fig. 2.2(a), $k = 2$) is repeatedly added to itself in such a way that the seed configuration is regarded as a unit and in the new structure these units are arranged with respect to each other according to the same symmetry as the original units in the seed configuration. In the next stage the previous configuration is always looked at as the seed. The construction of Fig. 2.2(b) is based on division of the original object and it can be easily seen how the subsequent replacement of the squares with five smaller squares leads to a self-similar, scale-invariant structure.

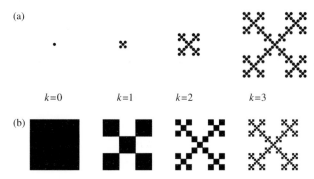

(a)

$k=0$ $k=1$ $k=2$ $k=3$

(b)

Fig. 2.2 (a) demonstrates how one can generate a growing fractal using an iteration procedure. In (b) an analogous structure is constructed by subsequent divisions of the original square. Both procedures lead to fractals for $k \to \infty$ with the same dimension $D \simeq 1.465$ [3].

One can generate many possible patterns by this technique; the fractal shown in Fig. 2.2 was chosen just because it has an open branching structure analogous to many observed biological fractals [3]. Only the first couple of steps (up to $k = 3$) of the construction are shown. *Mathematical fractals* are produced after an *infinite number of such iterations*. In this $k \to \infty$ limit the fractal displayed in Fig. 2.2(a) becomes infinitely large, while the details of Fig. 2.2(b) become so fine that the picture seems to 'evaporate' and cannot be seen any more. Our example shows a connected construction, but disconnected objects distributed in a non-trivial way in space can also form a fractal.

In any real system there is always a lower cutoff of the length scale; in our case this is represented by the size of the particles. In addition, a real object has a finite linear size which inevitably introduces an upper cutoff of the scale on which fractal scaling can be observed. This leads us to the conclusion that, in contrast to the mathematical fractals, for fractals observed *in natural phenomena (including biology) the anomalous scaling of the volume can be observed only between two well-defined length scales*.

Then, a possible definition for a biological fractal can be based on the requirement that a power-law scaling of $N(l)$ has to hold over *at least two orders of magnitude*.

2.1.2 Definitions

Because of the two main types of fractals demonstrated in Fig. 2.2, to define and determine the fractal dimension D one typically uses two related approaches.

For fractals having fixed L and details on a very small length scale, D is defined through the scaling of $N(l)$ as a function of decreasing l, where $N(l)$ is the number of d-dimensional balls of diameter l needed to cover the structure.

In the case of growing fractals, where there exists a smallest typical size a, one cuts out d-dimensional regions of linear size L from the object and the volume, $V(L)$, of the fractal within these regions is considered as a function of the linear size L of the object. When determining $V(L)$, the structure is covered by balls or boxes of unit volume ($l = a = 1$ is usually assumed); therefore $V(L) = N(L)$, where $N(L)$ is the number of such balls.

The fact that an object is a mathematical fractal then means that $N(l)$ diverges as $l \to 0$ or $L \to \infty$, respectively, according to a non-integer exponent.

Correspondingly, for fractals having a finite size and infinitely small ramifications we have

$$N(l) \sim l^{-D} \tag{2.3}$$

with

$$D = \lim_{l \to 0} \frac{\ln N(l)}{\ln(1/l)}, \tag{2.4}$$

while

$$N(L) \sim L^D \tag{2.5}$$

and

$$D = \lim_{L \to \infty} \frac{\ln N(L)}{\ln(L)}, \qquad (2.6)$$

for the *growing case*, where $l = 1$. Here, as well as in the following expressions, the symbol $\sim$ means that the proportionality factor, not written out in eqn (2.3), is independent of l.

Obviously, the above definitions for non-fractal objects give a trivial value for D coinciding with the embedding Euclidean dimension d. For example, the area (corresponding to the volume $V(L)$ in $d = 2$) of a circle grows as its squared radius, which according to eqn (2.6) results in $D = 2$.

Now we are in the position to calculate the dimension of the objects shown in Fig. 2.2. It is evident from the figure that for the growing case

$$N(L) = 5^k \qquad \text{with } L = 3^k, \qquad (2.7)$$

where k is the number of iterations completed. From here using eqn (2.6) we get the value $D = \ln 5 / \ln 3 = 1.465\ldots$, which is a number between $d = 1$ and $d = 2$ just as we expected.

2.1.3 Useful rules

In this section we mention a few rules which can be useful in predicting various properties related to the fractal structure of an object. Of course, because of the great variety of self-similar geometries the number of possible exceptions is not small and the rules listed below should be regarded, at least in part, as starting points for more accurate conclusions.

1. Many times it is the *projection* of a fractal which is of interest or can be experimentally studied (e.g. a picture of a fractal embedded into $d = 3$). In general, projecting a $D < d_s$-dimensional fractal onto a d_s-dimensional surface results in a structure with the same fractal dimension $D_p = D$. For $D \geq d_s$ the projection fills the surface, $D_p = d_s$.

2. It follows from (a) that for $D < d_s$ the density correlations $c(r)$ (see next section) within the projected image decay as a power law with an exponent $d_s - D$ instead of $d - D$ which is the exponent characterizing the algebraic decay of $c(r)$ in d.

3. Cutting out a d_s-dimensional slice (*cross-section*) of a D-dimensional fractal embedded into a d-dimensional space usually leads to a $(D+d_s-d)$-dimensional object. This seems to be true for self-affine fractals (next section) as well, with D being their local dimension.

4. Consider two sets A and B having fractal dimensions D_A and D_B, respectively. *Multiplying* them together results in a fractal with $D = D_A + D_B$. As a simple example, imagine a fractal which is made of parallel sticks arranged in such a

way that its cross-section is the fractal shown in Fig. 2.2(b). The dimension of this object is $D = 1 + \ln 5/\ln 3$.

5. The *union* of two fractal sets A and B with $D_A > D_B$ has the dimension $D = D_A$.

6. The fractal dimension of the *intersection* of two fractals with D_A and D_B is given by $D_{A \cap B} = D_A + D_B - d$. To see this, consider a box of linear size L within the overlapping region of two growing stochastic fractals. The density of A and B particles is respectively proportional to L^{D_A}/L^d and L^{D_B}/L^d. The number of overlapping sites $N \sim L^{D_{A \cap B}}$ is proportional to these densities and to the volume of the box which leads to the above relation. The rule concerning intersections of fractals with smooth hypersurfaces (rule (c)) is a special case of the present one.

7. The distribution of empty regions (holes) in a fractal of dimension D scales as a function of their linear size with an exponent $-D - 1$.

Self-similarity can be directly checked for a deterministic fractal constructed by iteration, but in the case of random structures one needs other methods to detect the fractal character of a given object. In fact, *random fractals are self-similar only in a statistical sense* (not exactly), and to describe them it is more appropriate to use the term scale invariance than self-similarity. Naturally, for demonstrating the presence of fractal scaling one can use the definition based on covering the given structure with balls of varying radii; however, this would be a rather troublesome procedure. It is more effective to calculate the so called *density–density or pair correlation function*

$$c(r) = \frac{1}{V} \sum_{r'} \rho(r + r')\, \rho(r'), \qquad (2.8)$$

which is the expectation value of the event that two points separated by r belong to the structure. For growing fractals the volume of the object is $V = N$, where N is the number of particles in the cluster, and eqn (2.9) gives the probability of finding a particle at the position $r + r'$, if there is one at r'. In eqn (2.9) ρ is the local density, i.e., $\rho(r) = 1$ if the point r belongs to the object, otherwise it is equal to zero. Ordinary fractals are typically isotropic (the correlations are not dependent on the direction) which means that the density correlations depend only on the distance r, so that $c(r) = c(r)$.

Now we can use the pair correlation function introduced above as a criterion for fractal geometry. An object is non-trivially scale-invariant if its correlation function determined according to eqn (2.9) is unchanged up to a constant under rescaling of lengths by an arbitrary factor b:

$$c(br) \sim b^{-\alpha} c(r), \qquad (2.9)$$

with α a non-integer number larger than zero and less than d. It can be shown that the only function which satisfies eqn (2.9) is the power-law dependence of $c(r)$ on r

$$c(r) \sim r^{-\alpha}, \qquad (2.10)$$

corresponding to an algebraic decay of the local density within a random fractal, since the pair correlation function is proportional to the density distribution around a given point. Let us calculate the number of particles $N(L)$ within a sphere of radius L from their density distribution

$$N(L) \sim \int_0^L c(r)d^d r \sim L^{d-\alpha}, \tag{2.11}$$

where the summation in eqn (2.8) has been replaced by integration. Comparing eqn (2.11) with eqn (2.5) we arrive at the relation

$$D = d - \alpha, \tag{2.12}$$

which is a result widely used for the determination of D from the density correlations within a random fractal.

2.1.4 Self-similar and self-affine fractals

There are three major types of fractals as regards their scaling behaviour. Self-similar fractals are invariant under isotropic rescaling of the coordinates, while for self-affine fractals scale invariance holds for affine (anisotropic) transformation. Until this point mainly the former case has been discussed.

The random motion of a particle represents a particularly simple example of stochastic processes leading to growing fractal structures. A widely studied case is when the particle undergoes a random walk (Brownian or diffusional motion) making steps of length distributed according to a Gaussian distribution in randomly selected directions. Such processes can be described in terms of the mean-squared distance $R^2 = \langle R^2(t) \rangle$ made by the particles during a given time interval t. For random walks, $R^2 \sim t$ independently of d, which means that the Brownian trajectory is a random fractal in spaces with $d > 2$. Indeed, measuring the volume of the trajectory by the total number of places visited by the particle making t steps, $(N(R) \sim t)$, the above expression is equivalent to

$$N(R) \sim R^2, \tag{2.13}$$

and comparing eqn (2.13) with eqn (2.5) we conclude that for random walks $D = 2 < d$ if $d > 2$. In this case, rather unusually, the fractal dimension is an integer number. However, the fact that it is definitely smaller than the embedding dimension indicates that the object must be non-trivially scale invariant.

In many physically relevant cases the structure of the objects is such that it is invariant under dilation transformation only if the lengths are rescaled by direction dependent factors. These anisotropic fractals are called self-affine [4–6].

Single-valued, nowhere-differentiable functions represent a simple and typical form in which self-affine fractals appear. If such a function $F(x)$ has the property

$$F(x) \simeq b^{-H} F(bx), \tag{2.14}$$

it is self-affine, where $H > 0$ is some exponent. Equation (2.14) expresses the fact that the function is invariant under the following rescaling: shrinking along the x axis by a factor $1/b$, followed by rescaling of values of the function (measured in a direction perpendicular to the direction in which the argument is changed) by a different factor equal to b^{-H}. In other words, by shrinking the function using the appropriate direction-dependent factors, it is rescaled onto itself. For some deterministic self-affine functions this can be done exactly, while for random functions the above considerations are valid in a stochastic sense (expressed by using the sign $\simeq$).

A definition of self-affinity equivalent to eqn (2.14) is given by the expression for the *height correlation function* $c(\Delta x)$:

$$c(\Delta x) = \langle [F(x + \Delta x) - F(x)]^2 \rangle \sim \Delta x^{2H}, \qquad (2.15)$$

which can be easily used for the determination of the exponent H. In addition to functions satisfying eqns (2.14) and (2.15), there are also self-affine fractals different from single-valued functions.

Let us first construct a deterministic self-affine model, in order to have an object which can be treated exactly.

An actual construction of such a bounded self-affine function on the unit interval is demonstrated in Fig. 2.3. The object is generated by a recursive procedure by replacing the intervals of the previous configuration with the generator having the form of an asymmetric and rotated letter z made of four intervals. However, the replacement this time should be done in a manner different from the earlier practice. Here every interval is regarded as a diagonal of a rectangle becoming increasingly elongated during the iteration. The basis of the rectangle is divided into four equal parts and the z-shaped generator replaces the diagonal in such a way that its turnovers are always at analogous positions (at the first quarter and the middle of the basis). The function becomes self-affine in the $k \to \infty$ limit.

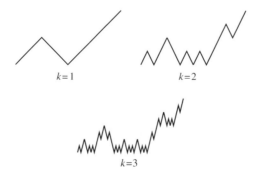

$k=1$ $k=2$

$k=3$

Fig. 2.3 Self-affine functions can be generated by iteration procedures. The single-valued character of the function is preserved by an appropriate distortion of the z-shaped generator ($k = 1$) of the structure [3].

Such a random function is, for example, the plot of the distances $X(t)$, measured from the origin as a function of time t, of a Brownian particle diffusing in one dimension. It is obvious that a so-called *fractional Brownian plot* for which $\langle X_H^2(t) \rangle \sim t^{2H}$ satisfies eqn (2.15).

Next we give a further basic feature of fractional Brownian motion. Calculating the Fourier spectrum of a fractional Brownian function one finds that the coefficients of the series, $A(f)$, are independent Gaussian random variables and their absolute value scales with the frequency f according to a power law as

$$|A(f)| \sim f^{-H-1/2}. \tag{2.16}$$

2.1.5 Multifractals

In the previous sections complex geometrical structures were discussed which could be interpreted in terms of a single fractal dimension. The present section is mainly concerned with the development of a formalism for the description of the situation when a singular distribution is defined on a fractal [7, 8].

It is typical for a large class of phenomena in nature that the behaviour of a system is determined by the spatial distribution of a scalar quantity, e.g., concentration, electric potential, probability, etc. For simpler geometries this distribution function and its derivatives are relatively smooth, and they usually contain only a few (or none) singularities, where the word singular corresponds to a local power-law behaviour of the function. (In other words, we call a function singular in the region surrounding point x if its local integral diverges or vanishes with a non-integer exponent when the region of integration goes to zero). In the case of fractals the situation is quite different: a process in nature involving a fractal may lead to a spatial distribution of the relevant quantities which possesses infinitely many singularities.

As an example, consider an isolated, charged object. If this object has sharp tips, the electric field around these tips becomes very large in accordance with the behaviour of the solution of the Laplace equation for the potential. In the case of charging the branching fractals produced in the $k \to \infty$ limit of constructions shown in Fig. 2.1 or 2.2 one has an infinite number of tips and corresponding singularities of the electric field. Moreover, the tips being at different positions, in general they have different local environments (configuration of the object in the region surrounding the given tip) which affect the strength of the singularity associated with that position.

The above-discussed time-independent distributions defined on a fractal substrate are called *fractal measures*. In general, a fractal measure possesses an infinite number of singularities of infinitely many types. The term '*multifractality*' expresses the fact that points corresponding to a given type of singularity typically form a fractal subset whose dimension depends on the type of singularity. The description of the multifractal formalism goes beyond the scope of the present section, but can be found, for example, in [3].

2.1.6 Methods for determining fractal dimensions

When one tries to determine the fractal dimension of biological structures in practice, it usually turns out that the direct application of definitions for D given in the previous sections is ineffective or cannot be accomplished. Instead, one is led to measure or calculate quantities which can be shown to be related to the fractal dimension of the objects. Three main approaches are used for the determination of these quantities: experimental, computational and theoretical.

Experimental methods for measuring fractal dimensions

A number of experimental techniques have been used to measure the fractal dimension of scale-invariant structures grown in various experiments. The most widely applied methods can be divided into the following categories: (a) digital image processing of two-dimensional pictures, (b) scattering experiments and (c) direct measurement of dimension-dependent physical properties.

(a) Digitizing the image of a fractal object is a standard way of obtaining quantitative data about geometrical shapes. The information is picked up by a scanner or an ordinary video camera and transmitted into the memory of a computer (typically a PC). The data are stored in the form of a two-dimensional array of pixels whose non-zero (equal to zero) elements correspond to regions occupied (not occupied) by the image. Once they are in the computer, the data can be evaluated using the methods described in the next section, where calculation of D for computer-generated clusters is discussed.

The only principal question related to processing of pictures arises if two-dimensional images of objects embedded into three dimensions are considered. It has already been mentioned that the fractal dimension of the projection of an object onto a $(d - m)$-dimensional plane is the same as its original fractal dimension, if $D < d - m$.

(b) Scattering experiments represent a powerful method of measuring the fractal dimension of structures. Depending on the characteristic length scales associated with the object to be studied, light, X-ray or neutron scattering can be used to reveal fractal properties. There are a number of ways to carry out a scattering experiment. One can investigate (i) the structure factor of a single fractal object, (ii) scattering by many clusters growing in time, (iii) the scattered beam from a fractal surface, etc.

Evaluation of numerical data

Throughout this section we assume that the information about the stochastic structures is stored in the form of d-dimensional arrays which correspond to the values of a function given at the nodes (or sites) of some underlying lattice. In the case of studying geometrical scaling only, the value of the function attributed to a point with given coordinates (the point being defined through the indexes of the array) is either 1 (the point belongs to the fractal) or 0 (the site is empty). When multifractal properties are investigated, the site function takes on arbitrary values. In general, such discrete sets of numbers are obtained by two main methods: (i) by digitizing pictures taken from objects produced in experiments and (ii) by numerical procedures used for the simulation of various biological structures. For convenience, in the following we shall frequently use the terminology 'particle' for a lattice site which belongs to the fractal (is filled) and cluster for the objects made of connected particles.

Below we discuss how to measure D for a single object. To make the estimates more accurate one usually calculates the fractal dimension for *many clusters* and averages over the results.

Perhaps the most practical method is to determine the number of particles $N(R) = R^D$ within a region of linear size R and obtain the fractal dimension D from the slopes of the plots of $\ln N(R)$ versus $\ln R$. If the centres of the regions of radius R are the particles of the cluster, than $N(R)$ is equivalent to the integral of the density correlation function. In practice one chooses a subset of randomly selected particles of the fractal (as many as needed for reasonable statistics) and determines $\langle N(R) \rangle$ for a sequence of growing R (or counts the number of particles in boxes of linear size L). In order to avoid undesirable effects caused by anomalous contributions appearing at the edge of the cluster one should not choose particles as centres close to the boundary region. The situation is shown in Fig. 2.4. Typically there is a deviation from scaling for small and large scales.

The roughness exponent H corresponding to *self-affine* fractals is usually determined from the definition of eqn (2.15). An alternative method is to investigate the scaling of the standard deviation $\sigma(l) = \left[\langle F^2(x) \rangle_x - \langle F(x) \rangle_x^2 \right]^{1/2}$ of the self-affine function F:

$$\langle \sigma(l) \rangle \sim l^H, \tag{2.17}$$

where the left-hand side is the average of the standard deviation of the function F calculated for regions of linear size l. The roughness exponent H can be calculated

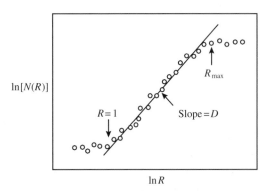

Fig. 2.4 Schematic log-log plot of the numerically determined number of particles $N(R)$ belonging to a fractal and being within a sphere of radius R. If R is smaller than the particle size or larger than the linear size of the structure, a trivial behaviour is observed. The fractal dimension is obtained by fitting a straight line to the data in the scaling region [3].

by determining $\sigma(l)$ for parts of the interfaces for various l. An averaging should be made over the segments of the same length and the results plotted on a double logarithmic plot as a function of l.

2.2 Stochastic processes

I. Derényi

2.2.1 The physics of microscopic objects

Every biological process eventually takes place at the molecular level. The physics of this microscopic realm is fundamentally different from the physics of our macroscopic world, and requires a completely different description. First of all, as the length-scale and velocity-scale go down to molecular scales, the Reynolds number goes down too, and we approach the overdamped regime in which inertia plays no role anymore [9] and where the velocity (and not the acceleration) of the objects is proportional to the forces acting on them. Secondly, there is Brownian motion. Microscopic objects are randomly kicked around by molecules of the surrounding medium, and the processes have an inherently stochastic nature.

Since the timescale of macroscopic processes is set by the velocity and acceleration of massive objects, the thermal fluctuations are negligible, and when we design a macroscopic device we try to suppress any stochastic element as much as possible. On the other hand, for microscopic objects every degree of freedom has inevitably a significant thermal energy $\frac{1}{2}k_B T$ on average (where T denotes the absolute temperature and k_B is the Boltzmann constant). The timing of the processes is therefore set by thermally assisted events (such as diffusion or activated transitions over energy barriers), and for the design of microscopic devices thermal fluctuations should be exploited rather than suppressed.

In general, the motion of any object in a thermal environment can be described by the *Langevin equation* [10]:

$$m\ddot{x}(t) = -\gamma \dot{x}(t) + \gamma \sqrt{2D}\xi(t) + F(x,t) , \qquad (2.18)$$

where x, m, γ and D denote the position, mass, viscous friction coefficient and diffusion coefficient of the object, respectively. The three force terms on the right-hand side of the equation are the viscous friction by the medium, $\gamma \dot{x}(t)$, the thermal noise coming from the molecules of the medium, $\gamma \sqrt{2D}\xi(t)$, and all the other forces unrelated to the medium, $F(x,t)$. Since the thermal noise term is a stochastic function, the Langevin equation is referred to as a stochastic differential equation. The noise factor, $\xi(t)$, is usually modelled by a Gaussian white noise with zero time average,

$$\langle \xi(t) \rangle = 0 , \qquad (2.19)$$

and autocorrelation function

$$\langle \xi(t)\xi(t') \rangle = \delta(t - t') . \tag{2.20}$$

Both the viscous friction and the thermal noise are exerted by the medium, and are not independent. Their magnitudes are connected by the fluctuation-dissipation theorem (or Einstein relation):

$$D = \frac{k_B T}{\gamma} . \tag{2.21}$$

For biomolecules in water solution the Langevin equation can be simplified. The ratio $\tau_{relax} = m/\gamma$ is a characteristic time scale of the Langevin equation (2.18), and tells us how long it takes for a particle to lose its initial velocity via viscous friction if the thermal noise and F are turned off. Multiplying this by the particle's characteristic velocity v, we get the characteristic distance $\lambda_{relax} = vm/\gamma$ in which the particle comes to a halt. Comparing this distance to the characteristic size of the particle a, we get some information about the strength of the viscous damping: if $\lambda_{relax}/a \ll 1$, the damping is strong, because the particle stops in a much shorter distance than its size; and if $\lambda_{relax}/a \gg 1$, the damping is weak. Supposing that m is proportional to $a^3\rho$ and γ is proportional to $a\eta$ (cf. Stokes law), where ρ and η are the density and dynamic viscosity of the medium, respectively, λ_{relax}/a becomes proportional to

$$R = \frac{va}{\eta/\rho} = \frac{va}{\nu} , \tag{2.22}$$

which is called the *Reynolds number* ($\nu = \eta/\rho$ is the kinematic viscosity). Thus, it is the Reynolds number that characterizes the strength of the damping. Low Reynolds number means strong damping.

Let us now estimate the Reynolds number for biological molecules. The typical size of a protein is of the order of nanometres ($a \approx 1$ nm), and the density and dynamic viscosity of water are $\rho \approx 10^3$ kg/m^3 and $\eta \approx 10^{-3}$ kg/s/m, respectively. The maximal forces acting on a protein are of the order of piconewtons (a few $k_B T$ over a few nanometres); thus the characteristic velocity of a protein cannot be much larger than $v \approx 1$ pN/$(a\eta) \approx 1$ m/s. This shows that the Reynolds number for biomolecules is of the order of 10^{-3} or even smaller, i.e., we are in the strongly damped or *overdamped* regime. A value of $R = 10^{-3}$ is a somewhat shocking result. It means that the viscous friction can stop a protein in a distance ($\sim 10^{-3}$ nm) much shorter than the size of the atoms.

In this overdamped regime, when the forces change, the velocity of a particle relaxes so quickly (during τ_{relax}) and over such a small distance ($\lambda_{relax} \ll a$) that the acceleration term (the derivative of the velocity with respect to time) on the left-hand side of the Langevin equation (2.18) can be neglected:

$$\dot{x}(t) = F(x, t)/\gamma + \sqrt{2D}\xi(t) . \tag{2.23}$$

This kind of reduction is called adiabatic elimination of the fast variables [10, 11]. Since the motion of biomolecules can be well described by the overdamped Langevin equation, from now on we will use only this version of the equation, and also the term Langevin equation will always refer to its overdamped version, eqn (2.23).

From this stochastic ordinary differential equation one can derive a deterministic partial differential equation, the *Fokker–Planck equation* (or Smoluchowski equation) [10], which describes the time evolution of the probability density $P(x, t)$ of the position of the particle:

$$\partial_t P(x, t) = -\partial_x J(x, t) , \tag{2.24}$$

where

$$J(x, t) = \frac{F(x, t)}{\gamma} P(x, t) - \frac{k_B T}{\gamma} \partial_x P(x, t) \tag{2.25}$$

is the probability current of the particle. If the force field $F(x, t)$ is the negative gradient of a potential, $F(x, t) = -\partial_x V(x, t)$, the probability current can be written in the form

$$J(x, t) = -\frac{k_B T}{\gamma} e^{-V(x,t)/k_B T} \partial_x \left[e^{V(x,t)/k_B T} P(x, t) \right] . \tag{2.26}$$

2.2.2 Kramers formula and Arrhenius law

In many systems Brownian particles wiggle in deep potential wells (compared to $k_B T$) for long periods of time, rarely interrupted by quick jumps into one of the neighbouring wells. If the potential is static or changes in a much longer timescale than the duration of these jumps (which is usually the case), a *kinetic* approach can be used to describe the motion of the particles, with transition rate constants between discrete states.

The discipline of rate theory (for a review see [12]) was created when Arrhenius [13] extensively discussed various reaction-rate data and showed that they vary on a logarithmic scale linearly with the inverse temperature T^{-1}. In other words, the escape (or jumping) rate constants, k, follow the Arrhenius law

$$k = \nu e^{-\Delta E/k_B T}, \tag{2.27}$$

where ΔE denotes the threshold energy for activation and ν is a frequency prefactor.

Using Kramers' method [10, 14], the Arrhenius law can be easily derived from the Fokker–Planck equation for an overdamped Brownian particle moving in a one-dimensional potential $V(x)$ (depicted in Fig. 2.5), if the potential well from which the particle is trying to escape is much deeper than $k_B T$. In this case the probability density near the bottom can be well approximated by its equilibrium value

$$P_{eq}(x) = P_0 e^{-V(x)/k_B T} , \tag{2.28}$$

which can be derived from eqn (2.26) by setting its right-hand side to zero. The normalization factor is approximately

$$P_0 = \frac{1}{\int_b^c e^{-V(x)/k_B T} \, dx} \tag{2.29}$$

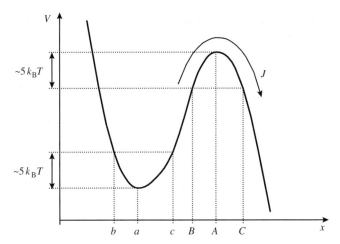

Fig. 2.5 Potential $V(x)$ with a deep well and a barrier over which an overdamped Brownian particle tries to escape.

because the vast majority of the probability falls into the interval $[b,c]$ where the potential difference from the bottom of the potential is not larger than a few (~ 5) $k_B T$. Kramers' approach is based on the assumption that the probability current over the potential barrier between B and C is a constant J. Indeed, this condition holds, because the interval $[B,C]$ contains only a very small fraction of the total probability. Another assumption is that the probability distribution for $x \geq C$ is zero, because the particle has basically no chance to get back to the well and the escape can be considered to be complete. Thus, after rearranging and integrating eqn (2.26) from B to C we get

$$J \frac{\gamma}{k_B T} \int_B^C e^{V(x)/k_B T} \, dx = - \left[e^{V(x)/k_B T} P_{eq}(x) \right]_{x=B}^C , \qquad (2.30)$$

where the expression between the square brackets is P_0 for $x = B$ and zero for $x = C$. From this, the current J over the barrier (which is equivalent to the escape rate constant k) can be expressed as

$$J \equiv k = \frac{D}{\int_b^c e^{-V(x)/k_B T} \, dx \int_B^C e^{V(x)/k_B T} \, dx}$$

$$= \frac{D e^{-[V(A)-V(a)]/k_B T}}{\int_b^c e^{-[V(x)-V(a)]/k_B T} \, dx \int_B^C e^{-[V(A)-V(x)]/k_B T} \, dx} . \qquad (2.31)$$

This expression has indeed the same form as that of the Arrhenius law, eqn (2.27). Here the activation energy ΔE is the height of the barrier $V(A) - V(a)$, and the frequency prefactor ν depends only on the shape of the potential near the bottom of the well and the top of the barrier.

This derivation holds even if the potential $V(x)$ changes in time but much more slowly than the intrawell relaxation time of the particle [15]. In this case the escape rate constant becomes also time dependent.

Most chemical reactions can also be described in terms of kinetic rate constants; so if they are present, they represent another source of stochasticity in molecular processes in addition to the thermal noise. The waiting time for any escape process or chemical reaction characterized by a rate constant k has an exponential distribution with mean value $1/k$.

2.3 Continuous phase transitions

Z. Csahók

In the following chapter we discuss the relevance of the so-called self-organized criticality (SOC) for biology. However, before describing this more advanced, non-equilibrium concept we give a short introduction to the closely related precursor, the second-order (or continuous) phase transition occurring in equilibrium.

Phase transitions can be easily understood at a simple thermodynamic level. Let us consider a substance which can exist in two different phases like water (liquid and ice) or iron (paramagnetic and ferromagnetic). Usually one of the phases is *disordered* while the other is *ordered*. The distinction is based on the *symmetry* of the state: the symmetric (or isotropic) state is the disordered one. To characterize the strength of ordering at given values of the thermodynamic parameters (like the temperature) we introduce an *order parameter*. It measures how well ordered the substance is. By convention if the order parameter is zero, we speak of a completely disordered state (e.g., liquid water), and if it is non-zero, then the substance is in an ordered state (e.g., ice crystal). As is known from thermodynamics, at constant temperature the phase with the lowest free energy is stable. Since the free energy generally can also be a function of the order parameter, this energy minimum requirement will select which phase can be observed at a given temperature T. In this case the temperature plays the role of a *control parameter* which allows for tuning the system to the phase transition.

In Fig. 2.6 we sketch two possible ways leading to a phase transition. When the temperature is above the *critical temperature* T_c (where the transition occurs), the only globally stable solution to the energy minimum criterion is the phase with zero order parameter ($m = 0$). This means that for $T > T_c$ the system is in its disordered state. Figure 2.6(a) shows a case as the temperature is lowered, where a non-trivial ($m > 0$) minimum first appears at $T = T_c$ and then shifts to larger values as the temperature is lowered (see inset in Fig. 2.6(a)). This is the scenario of a *first-order* phase transition. The characteristic feature of this type of transition is a jump discontinuity in the order parameter (and certain other quantities). In contrast, Fig. 2.6(b) demonstrates another type of transition where the order parameter changes *continuously* as the temperature is lowered below T_c. This type of transition is referred as a *second-order*

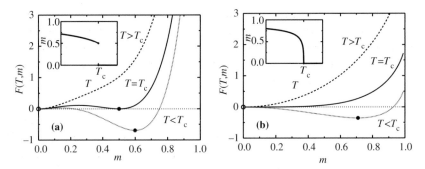

Fig. 2.6 Two routes to a phase transition: (a) first order and (b) second order. The insets show the order parameter as a function of the temperature.

phase transition and it has received much larger attention in the past decades. A motivation for this interest lies in the special properties of such transitions.

Without loss of generality let us consider a more specific example for a second-order transition: a magnetic material at temperature T and magnetic field H. If this system shows a paramagnetic–ferromagnetic second-order transition at T_c (for $H = 0$), then it is convenient to use the *reduced temperature*

$$t = \frac{T - T_c}{T_c} \tag{2.32}$$

as the control parameter instead of T.

Measuring the physical properties of the sample reveals divergences of various physical observables as the critical temperature is approached. The order parameter in this case is the zero-field ($H = 0$) magnetization M_0, since it is zero if $T > T_c$ and non-zero below T_c. In the vicinity of T_c it behaves as

$$M_0(t) \sim |t|^\beta, \tag{2.33}$$

where β is the *critical exponent* of the magnetization. Similarly, for the specific heat (which gives the change of energy for a small change of temperature)

$$C_{H=0}(t) \sim |t|^{-\alpha}, \tag{2.34}$$

and the susceptibility (which is the sensitivity of the magnetization with respect to the external field H) scales as

$$\chi(t) \sim |t|^{-\gamma}. \tag{2.35}$$

Both of these quantities describe a response of the system to some external perturbation. Close to T_c they diverge, showing that there the system is extremely sensitive: it is in a critical state.

The above-defined three critical exponents (α, β, γ) are not independent of each other. A more detailed analysis shows that the *exponent relation*

$$\beta = \frac{2 - \alpha - \gamma}{2} \tag{2.36}$$

holds.

Near the transition point the spontaneous fluctuations in the system become large due to the high susceptibilities. For the case of a fluid these strong fluctuations are observable as the decrease of light transmittance (*critical opalescence*). Since the length scale ξ associated with these fluctuations, i.e., the typical size of fluid droplets, also has a power-law divergence

$$\xi \sim t^{-\nu}, \tag{2.37}$$

at the critical point there will be no typical length scale except the trivial lower (atomic size) and upper (system size) scales. This fact is manifested via the fractal [3] structure of the fluctuations in the system and it is closely connected to other power-law divergences present at T_c. The fractality from the experimenter's point of view means that the fluctuations are statistically invariant under the transformation

$$(x, y, \ldots) \longmapsto (\lambda x, \lambda y, \ldots), \tag{2.38}$$

i.e., no typical length scale can be identified.

2.3.1 The Potts model

Using the thermodynamic approach it is possible to derive the critical exponents only heuristically, based on symmetry arguments, supposing some form of the (coarse grained) free energy near the critical point. This method is used by the Landau theory of critical phenomena. To overcome the limitations of this approach one has to introduce models which include more details about the interactions leading to the phase transition. A number of different models can be constructed depending on the level of abstraction at which interactions are handled. Here we discuss a rather general lattice model introduced by Potts in 1952 [16].

The q-state Potts model consists of a set of 'spins' (or particles) $\{s_i\}$, each of which may have integer values $s_i = 0, 1, \ldots (q - 1)$. These spins sit on a lattice and the Hamiltonian (the energy function) is defined as

$$H[\{s_i\}] = -\sum_{\langle ij \rangle} J \, \delta_{\mathrm{Kr}}(s_i, s_j), \tag{2.39}$$

where δ_{Kr} is the Kronecker delta function and the summation goes over nearest neighbours only (short-range interaction). The meaning of the energy function eqn (2.39) is that only particles with $s_i = s_j$ 'like' each other, only such combinations lower the energy of the system as illustrated in Fig. 2.7.

If the temperature is high then all the q states will be equally populated, so any quantity can serve as an order parameter which measures the difference of the distribution of spin states from uniform. At low temperatures the system organizes itself into a configuration where most spins are in a randomly selected state while the other states are weakly populated.

The Potts model is related to many other lattice models in statistical physics [17]. For $q = 2$ (two states: 'up' and 'down') it is equivalent to the well-known Ising

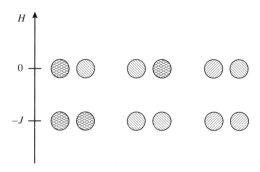

Fig. 2.7 Interaction energy of particles in the Potts model.

model. The $q = 1$ limit reproduces the percolation problem, while $q = 0$ can be mapped to the resistor network problem.

2.3.2 Mean-field approximation

Although some of the special cases mentioned above can be solved exactly, no general solution exists to the Potts model itself. Here we present the simplest approach to determine the exponents [17]. Consider the following slightly modified Hamiltonian

$$H[\{s_i\}] = -\frac{zJ}{N} \sum_{i<j} \delta_{Kr}(s_i, s_j), \tag{2.40}$$

where z is the coordination number of the lattice (number of neighbours, i.e., $z = 4$ for a planar square lattice) and N is the total number of spins. In contrast to eqn (2.39) this Hamiltonian allows long-range interactions, since the interaction of every spin-pair contributes to the total energy. In other words, a spin is affected not only by its neighbours, but rather by the mean state of the whole system. For this reason such an approximation is called a *mean-field* description. This approach has several deficiencies: by introducing long-range interactions it neglects the fluctuations which are essential parts of phase transitions, and therefore does not give correct results for the critical temperature and the exponents. Nevertheless, since it gives a qualitatively correct picture, it is worth examining.

Instead of accounting for the state of every spin we are going to consider the distribution of spins by states. Let x_i denote the fraction of spins that are in spin state $i = 0, 1, \ldots (q-1)$. The probability of the Kronecker delta in eqn (2.40) being 1 is

$$\Pr[\delta_{Kr}(s_i, s_j) = 1] = \sum_{i=0}^{q-1} x_i^2. \tag{2.41}$$

Then the energy per spin is

$$e = \frac{E}{N} = -\frac{zJ}{N^2} \frac{N^2}{2} \sum_{i=0}^{q-1} x_i^2 = -\frac{1}{2} zJ \sum_{i=0}^{q-1} x_i^2 \tag{2.42}$$

and the entropy per spin is given by

$$s = \frac{S}{N} = -k_B \sum_{i=0}^{q-1} x_i \ln x_i. \tag{2.43}$$

The free energy is then

$$f = e - Ts = -\tfrac{1}{2}zJ \sum_i x_i^2 + k_B T \sum_i x_i \ln x_i. \tag{2.44}$$

Supposing that in the ordered phase the $i = 0$ state will be populated, we look for a solution in the form of

$$x_0 = \frac{1}{q} + \frac{q-1}{q}m \tag{2.45}$$

and

$$x_i = \frac{1}{q} - \frac{1}{q}m, \tag{2.46}$$

where the value of the order parameter $0 \le m \le 1$ is to be chosen so as to minimize the free energy, eqn (2.44). In the disordered phase $m = 0$, since all sites are equally populated, while in the ordered state $m > 0$. The choice eqns (2.45) and (2.46) trivially satisfies the normalization condition $\sum_i x_i = 1$. Using eqns (2.44), (2.45), and (2.46) one finds that

$$\frac{f}{k_B T} = \frac{1 + (q-1)m}{q} \ln[1 + (q-1)m] + \frac{q-1}{q}(1-m)\ln(1-m)$$
$$- K\frac{q-1}{2q}m^2, \tag{2.47}$$

with $K = zJ/(k_B T)$. The equilibrium order parameter m_0 as a function of temperature T has to be determined from $f'(m_0) = 0$. It is easily seen that $m_0 = 0$ is always a solution, but at sufficiently low temperatures other solutions with $m_0 > 0$ emerge which may lead a lower free energy. The critical temperature T_c is then defined as the temperature at which this shift of the absolute minimum of the free energy occurs.

It is instructive to expand eqn (2.47) in powers of m around $m = 0$. Up to the third order one gets

$$\frac{f}{k_B T} = \frac{q-1}{2q}(q-K)m^2 - \frac{1}{6}(q-1)(q-2)m^3 + \cdots. \tag{2.48}$$

For $q > 2$ the coefficient in the cubic term becomes negative, which indicates a possibly first-order transition. If $q < 2$, then the transition in this mean-field limit is second order (cf. Fig. 2.6(b)). The critical exponent for the order parameter obtained by this method is

$$\beta = \tfrac{1}{2}.$$

As we have already mentioned above, generally the results of such a mean-field calculation should not be regarded as the solution of the model. In fact, the nature of

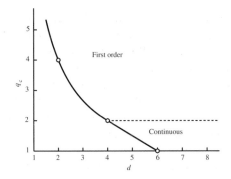

Fig. 2.8 Schematic plot of $q_c(d)$, the critical value of q beyond which the transition is mean-field-like (first order for $q > 2$ and continuous for $q \leq 2$). The known points $q_c(2) = 4$, $q_c(4) = 2$ and $q_c(6) = 1$ are denoted by open circles (after [17]).

the transition depends also on the dimension d of the underlying lattice. The dimensionality together with the lattice topology (i.e., square, triangular, etc.) determines the number of neighbouring sites z. With increasing lattice dimension a site gets more neighbours and thus the system approaches the mean-field limit where the lattice is fully connected. Therefore, it is expected that above a certain *critical dimension* d_c (or conversely, above some critical value of q) the mean-field behaviour will be exact (Fig. 2.8).

References

[1] B. B. Mandelbrot. *The Fractal Geometry of Nature*. Freeman, San Francisco, 1982.

[2] T. A. Witten and L. M. Sander. Diffusion-limited aggregation, a kinetic critical phenomenon. *Phys. Rev.* **B27**, 5686, 1983.

[3] T. Vicsek. *Fractal Growth Phenomena*. World Scientific, Singapore, 1992.

[4] F. Family and T. Vicsek (ed.). *Dynamics of Fractal Surfaces*. World Scientific, Singapore, 1991.

[5] B. B. Mandelbrot. Self-affine fractals sets. In *Fractals in Physics* (ed. L. Pietronero and E. Tosatti). Elsevier, Amsterdam, 1986, pp. 3–28.

[6] J. Kertész and T. Vicsek. Self-affine interfaces. In *Fractals in Sciences* (ed. A. Bunde and S. Havlin). Springer, Berlin, 1994, pp. 89–116.

[7] T. C. Halsey, M. H. Jensen, L. P. Kadanoff, I. Procacia and B. I. Shraiman. Fractal measures and their singularities: the characterization of strange sets. *Phys. Rev.* A33: 1141–1151, 1986.

[8] B. B. Mandelbrot. An introduction to multifractal distribution functions. In *Random Fluctuations and Pattern Growth* (ed. Stanley, H. E. and Ostrowsky, N.). Kluwer, Dordrecht, 1988, pp. 279–291.

[9] E. M. Purcell. Life at low Reynolds number. *Am. J. Phys.* 45: 3–11, 1977.

[10] H. Risken. *The Fokker–Planck Equation*. Springer-Verlag, Berlin, 1989.

[11] H. Haken. *Synergetics, An Introduction*. Springer-Verlag, Berlin, 1983.

[12] P. Hänggi, P. Talkner and M. Borkovec. Reaction-rate theory: fifty years after Kramers. *Rev. Mod. Phys.* 62: 251–341, 1990.

[13] S. Arrhenius. Über die Reaktionsgeschwindigkeit bei der Inversion von Rhrzucker durch Säuren. *Z. Phys. Chem.* 4: 226–248, 1889.

[14] H. A. Kramers. Brownian motion in a field of force and the diffusion model of chemical reactions. *Physica* 7: 284–304, 1940.

[15] I. Derényi and R. D. Astumian. Intrawell relaxation time: the limit of the adiabatic approximation. *Phys. Rev. Lett.* 82: 2623–2627, 1999.

[16] R. B. Potts. Some generalized order-disorder transformations. *Proc. Camb. Phil. Soc.* 48: 106, 1952.

[17] F. Y. Wu. The Potts model. *Rev. Mod. Phys.* 54: 235, 1982.

3 Self-organized criticality (SOC)

Z. Csahók

Perhaps the most conspicuous feature of life is its complexity. This complexity is due to processes capable of generating elaborate structures from such relatively simple building blocks as atoms. On the other hand, we know that the laws of physics (or chemistry) are rather simple and do not contain *a priori* information about highly organized patterns. Therefore, there must be some simple mechanisms which lead to complex spatial and temporal behaviour. Since the resulting structures appear in nature without any external enforcement, purely due to the interaction of the units in a system, we call the process of generating complexity *self-organization*.

In thermodynamics we define two distinct classes of systems. Closed systems left alone for a very long time approach the so-called thermal equilibrium state in which the temperature is constant everywhere and no macroscopic changes take place in time. In contrast, in non-equilibrium systems there is motion even in the so-called stationary states when the overall characteristics of the system do not change in time. It is in the non-equilibrium state that structures can emerge spontaneously from an originally homogeneous medium.

A specific kind of complexity can be observed in systems slowly driven to a stationary state. For example, slow driving means the gradual addition of energy to units of a system which may also lose this energy due to their interaction. In many cases this stationary state may represent a delicate balance among the units of the system. If the interaction between two units is such that a change exceeding a critical value in one unit results in a similar change in the neighbouring unit, than large avalanche-like series of changes may take place in the system when it is close to a critical (balanced) state. The system becomes very sensitive to perturbations or, in other words, critical.

As will be discussed later several biological systems can be considered both self-organized and critical. In the present chapter we describe the major concepts related to *self-organized criticality* (SOC) and present a few examples from biology which may be best interpreted in terms of SOC.

3.1 SOC model

So far we have considered systems in which the critical state was reached by tuning some parameter, typically the temperature. There are, however, systems in nature that exist in a critical state *without* apparent tunable parameters. Such systems include granular materials, earthquakes, fracture, river networks, complex biological systems (e.g., the brain, a collection of competing species), and economic systems. Systems which can reach their critical state without parameter tuning are termed *self-organizing critical* (SOC). The concept of SOC has been widely used in the studies of complex systems.

The standard model for SOC was introduced by Bak, Tang and Wiesenfeld in 1987 [1]. To provide an intuitive insight into the phenomenon, they used a sandpile as an example. Consider the typical childhood experience of building a sandpile by depositing sand on a surface. At first, sand is accumulated, making a sandpile. As one adds more and more sand to the pile, eventually at some place the slope of the sandpile will reach its critical value. Adding one more grain to that site will trigger an avalanche. Sand will then be redistributed and thus may create avalanches at various other places in the pile. Avalanches may actually span the whole sandpile.

This phenomenon is modelled by a discrete variable $h(\mathbf{r})$ defined on a d-dimensional lattice. For the sake of simplicity we are going to consider here only the $d = 2$ square-lattice case. The rules for the evolution of the system are as follows.

Step 1 Choose a site (x, y) at random;
Step 2 Add one 'grain' to that site: $h(x, y) \leftarrow h(x, y) + 1$;
Step 3 If $h(x, y) \leq 4$, then continue on Step 1.
Step 4 The pile is too 'steep' locally, so sand is redistributed among the nearest neighbours:

$$h(x, y) \leftarrow h(x, y) - 4$$

and

$$h(x', y') \leftarrow h(x', y') + 1,$$

where (x', y') denotes the nearest neighbours of the site (x, y).
Step 5 The previous step is repeated until no site with $h > 4$ is found and then Step 1 is performed again.

Step 1 and Step 2 represent a continuous uniform adiabatic (infinitely slow) driving process which makes the system far-from-equilibrium. Step 4 and Step 5 are called the process of relaxation. The size of an avalanche is defined by the number of times Step 4 was executed after adding a single grain. Large-scale computer simulations demonstrated that avalanches are power-law distributed:

$$P(\geq s) \sim s^{-\tau}, \tag{3.1}$$

where $P(\geq s)$ is the number of avalanches larger that s and the exponent τ is found to be around 0.25.

One should note that the role of the slope in this model is replaced by the height; this is why the sandpile picture is only a way to gain an intuitive gasp of the concept. Nevertheless, it is possible to construct models with critical slope instead of critical height, but this does not solve the problem of sandpiles, since they are *not* critical in reality [2]. It turns out that the behaviour of a sandpile is more reminiscent of a first-order transition due to the existence of not a single but two distinct critical slopes: the maximum angle of stability (θ_m) and the angle of repose ($\theta_r < \theta_m$). When the slope of the sand exceeds θ_m, an avalanche starts and it does not stop until the slope goes back to θ_r. The overall behaviour then will be a rather periodic appearance of avalanches and no power-law size distribution can be observed.

3.2 Applications in biology

The existence of large events that spread throughout the whole system can be generally considered as the forerunner of self-organized criticality. Such large-scale phenomena have been observed when tracing the extinction events during evolution. It turned out (see Fig. 3.1) that the extinction rate was not constant, as one would conjecture from a 'smooth' evolutionary process, but it showed peaks corresponding to major extinction events in our history. Although one can not rule out sudden environmental changes as the cause of mass extinction, it is also reasonable to consider evolution as a self-organized critical process.

Below we give some theoretical support to the idea of self-organized critical evolution and at the end of this section we describe another SOC system in biology: the lungs.

3.2.1 SOC model of evolution

In theoretical models of evolution the concept of *fitness* plays an essential role. The fitness of a species represents its ability to survive environmental changes as a function of its genetic code. There is no exact way to deduce a numerical value of fitness from a real DNA sequence, so this quantity is purely heuristic. Nevertheless, it allows us to construct models which grasp the main ingredients of evolution: mutation and selection. Within the framework of a model one can change the genetic code and

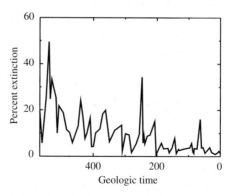

Fig. 3.1 Variation of the extinction rate during biological evolution (from [3]).

calculate the corresponding change in fitness (*mutation*) and then, if the new value of fitness is higher, accept the mutation or otherwise reject it (*selection*). Such a simple model would drive the system to a local maximum of fitness; therefore it is necessary to accept non-beneficial moves with some small probability. Nevertheless, the model described above would still lead to an equilibrium state which is not the case in real evolution, as we have discussed above.

In order to construct an SOC model for evolution [4, 5] let us consider an ecosystem of N species, where the fitness of the ith species ($i = 1, 2, \ldots, N$) is a real number

$$0 \leq B_i \leq 1. \tag{3.2}$$

Species with low fitness have larger selection pressure and therefore change their fitness more rapidly than those with higher fitness (for this reason fitness is denoted by B as *barrier*). For the sake of simplicity we implement a coarse-grained version of these dynamics:

1. We choose the species with the smallest fitness

$$B_j = \min_{i=1,\ldots,N} B_i. \tag{3.3}$$

2. Then this species makes an adaptive change (or mutation), which is represented by taking a random number for its fitness

$$B_j = \text{RND}, \tag{3.4}$$

where RND is equally distributed in [0,1].

3. Finally, the fitnesses of its two neighbours are also changed due to interaction

$$B_{j+1} = \text{RND}, \quad \text{and} \quad B_{j-1} = \text{RND}. \tag{3.5}$$

Taking only the first two steps would lead to a state where all species have the highest fitness ($B_i \equiv 1$). There would be no further dynamics possible in this state, and so it would correspond more to a 'dead' state then to a living and evolving ecology. The interaction step (step 3) is therefore essential; it corresponds, for example, to the fact that neighbouring species are all consecutive parts of the food chain.

This model can be easily implemented in a computer. Simulations are started from a random initial configuration of fitness. After some thousands of iterations the distribution of fitnesses tends to a non-trivial stable distribution (Fig. 3.2(a)). The distribution is flat below and above a threshold $B_c \sim 0.67$. It turns out that mutations take place through fitnesses that are smaller than B_c. This critical fitness is selected by the dynamics, not by tuning some external parameter. Therefore, we can say that the system has reached a self-organized state.

The activity in the systems can be displayed on a space–time plot of mutations (Fig. 3.2(b)). One can observe the intermittent nature of the evolution: after a period of localized mutations the position (i.e., the species affected by mutation) changes abruptly. In Fig. 3.2(c) we plotted the distribution of distances of consecutive

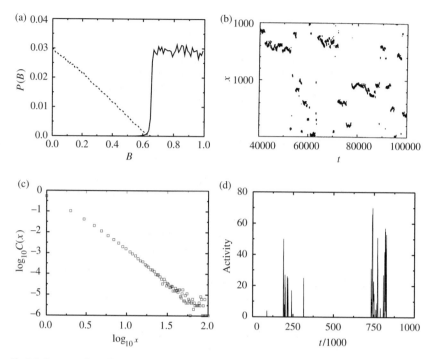

Fig. 3.2 Punctuated equilibrium behaviour in the SOC model for evolution. (a) Distribution of barriers (solid curve) and of active barriers (dashed curve) in the critical state. (b) Typical mutation activity of the system displayed in the space–time domain. (c) Distribution of distances between subsequent mutations. (d) Activity of a site averaged over a thousand consecutive iterations.

mutations which characterizes the correlations in the system. The straight line on the log–log plot indicates a power-law distribution, i.e., there is no typical correlation length and hence the system is critical.

It is instructive to plot the local activity of a site (in this case defined as the number of mutations within 1000 iterations) as a function of time (Fig. 3.2(d)). One can observe long periods of passivity interrupted by sudden bursts of activity. The temporal separation between bursts is even more pronounced if one takes into account that the real (physical) time needed for a change of barrier B scales as $\sim \exp(-B/\tau)$, where τ is the characteristic timescale for the mutations.

The model exhibits punctuated equilibrium: although the macroscopic features are stationary on a microscopic scale, large sequences of mutations (avalanches) can be observed. A straightforward way to define an avalanche in this system is to consider the number of subsequent mutations below a threshold B. With this definition there is a hierarchy of avalanches, each defined by their respective threshold. During an avalanche, sites may mutate more than once. For a threshold below the critical one ($B < B_c$) there is a typical scale of avalanches. This size diverges approaching B_c and one observes a power-law distribution of avalanche size ($P(\geq s) \sim s^{-\tau}$, where $\tau \simeq 1.07$ [6]), indicating that avalanches of all sizes are present, including catastrophic ones which span the whole system.

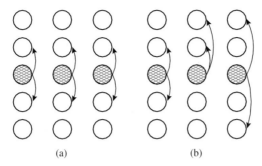

Fig. 3.3 Three consequent mutations of the same site (shaded): (a) in the original model the interaction is always nearest neighbour, (b) in the mean-field version the interaction partners are chosen randomly in each iteration.

In real evolution, large extinction events may be thought of as large avalanches caused by the intrinsic dynamics of biology. Of course, one can not exclude the effect of large cataclysmic events like volcanic eruptions or meteorites. Nevertheless, the model indicates that there may be no need for such events to explain an intermittent evolution: the ecology is able to self-organize itself into a critical state with avalanches of all sizes.

In the model there was nearest-neighbour interaction between the species. In order to formulate a solvable mean-field approximation [7] we consider *random neighbour* interaction, i.e., after having changed the barrier of a species, $K - 1$ other species are also assigned a random fitness. The main difference is that in this case the interacting sites are chosen randomly at every update rather than being fixed as in the original model. Therefore, the $K = 3$ case is not equivalent to the original model, since we are neglecting the correlations between neighbouring sites by reassigning interaction partners (Fig. 3.3). This redistribution of interactions validates the use of distribution functions for the barriers.

Let the distribution of barriers (x) in the ecology be $p(x, t)$. Then the smallest barrier is distributed as

$$p_1(x, t) = Np(x)Q^{N-1}(x, t) = -NQ^{N-1}\frac{dQ}{dx} = -\frac{dQ^N}{dx}, \qquad (3.6)$$

where

$$Q(x, t) = \int_x^1 dx' p(x', t). \qquad (3.7)$$

It is easy to show that

$$Q(0, t) = 1, \quad \text{and} \quad Q(1, t) = 0. \qquad (3.8)$$

The evolution equation for $p(x, t)$ is the following:

$$p(x, t+1) - p(x, t) = -\frac{1}{N}p_1(x, t) - \frac{K-1}{N-1}\left(p(x, t) - \frac{1}{N}p_1(x, t)\right) + \frac{K}{N}. \qquad (3.9)$$

The first term on the right-hand side of eqn (3.9) represents the removal of the smallest barrier value. The second term takes into account the removal of $K - 1$ of the $N - 1$

barrier values remaining after the smallest has been removed from the set of N values. The last term corresponds to the addition of K equally distributed random barriers replacing the removed ones.

We are looking for stationary solution $p(x)$ such that the left-hand side of eqn (3.9) vanishes. Substituting eqn (3.6) into eqn (3.9) we have

$$0 = -\frac{dQ^N}{dx}\left(\frac{1}{N} + \frac{K-1}{N(N-1)}\right) + \frac{K-1}{N-1}\frac{dQ}{dx} + \frac{K}{N}. \tag{3.10}$$

Integrating from x to 1 results in

$$0 = Q^N\left(\frac{1}{N} + \frac{K-1}{N(N-1)}\right) - \frac{K-1}{N-1}Q + \frac{K}{N}(1-x), \tag{3.11}$$

or

$$0 = Q^N(N-K) + N(K-1)Q - K(N-1)(1-x). \tag{3.12}$$

This is an equation for $Q(x)$ which cannot be generally solved exactly. However, it is possible to find approximate solutions in the limit $N \gg K$.

Let us now consider the two extremities when $Q \ll 1$ and $1 - Q \ll 1$, i.e., when $1 - x \ll 1$ and $x \ll 1$, respectively. The scale of Q (whether it can be considered small or large) is set by the relative significance the first and the second terms in eqn (3.12).

If Q is small, then the first term is irrelevant, that is

$$Q^N(N-K) \ll N(K-1)Q, \tag{3.13}$$

giving

$$Q^{N-1} \ll \frac{N(K-1)}{N-K} \approx K - 1 = O(1). \tag{3.14}$$

Thus from eqn (3.12) we have

$$Q_a = \frac{K(N-1)}{N(K-1)}(1-x) \approx \frac{K}{K-1}(1-x). \tag{3.15}$$

The criterion of eqn (3.14) and the solution Q_a are consistent only if

$$x - 1/K \gg O(1/N). \tag{3.16}$$

Conversely, if Q is large (close to 1), then the second term becomes negligible compared to the other terms. The solution in this case is given by

$$Q_b^N = \frac{-N(K-1)}{(N-K)} + \frac{K(N-1)}{N-K}(1-x) \approx 1 - Kx. \tag{3.17}$$

The consistency criterium now reads

$$1/K - x \gg O(1/N). \tag{3.18}$$

Using eqn (3.7) we have

$$p(x) = \begin{cases} K/N, & \text{if } x - 1/K \gg O(1/N); \\ K/(K-1), & \text{if } 1/K - x \gg O(1/N). \end{cases} \qquad (3.19)$$

In the limit $N \to \infty$ the distribution function of the barriers develops a discontinuity at $x = B_c = 1/K$. It vanishes below the threshold and is constant above it, in accord with the results of numerical simulations (cf. solid line in Fig 3.2(a)).

The above calculation does not give the correct value for B_c. From the simulations we know that it is around 0.67 but the calculations give for $K = 3$ the value of $1/3$. This discrepancy is due to the fact that in the calculation we have neglected the correlations. In fact, a more realistic approximation can be constructed by assuming that the two smallest barriers and a randomly selected third value are replaced in each time step.

3.2.2 SOC in lung inflation

The morphology of the mammalian lung has fascinated physiologists for over a century. Recent investigations [8] showed that bronchial airways have a fractal tree-like structure which can be approximated by a branching process, i.e., the number of branches in a 'generation' of the tree is about twice as large as in the previous generation (see Fig. 3.5(a)).

It is even more surprising that during inhalation the rate of air intake is not smoothly varying over time. In the actual experiment [9] an emptied dog lung was inflated by a constant flow and the terminal airway resistance (R_t) was measured. It was found that R_t decreased in discrete jumps (Fig. 3.4), which corresponds to a non-constant rate of inflation of the lung. Both the sizes of the jumps and the time intervals between them vary significantly within a single experiment. Magnifying a small portion of the curve one finds structures which are statistically similar to the original curve.

Therefore, the appropriate data analysis in this case is studying the distribution function of the sizes of the jumps and the time delays. To perform this analysis one needs a relatively large number of independent experiments (> 10) in order to obtain good statistics for the data. The results showed that the distribution functions are power-law over a range of more than two decades. We have seen previously that the presence of a power-law distribution indicates the lack of a characteristic scale: if there was a typical scale, then the distribution was exponential.

This result suggests the interpretation that the airways do not open independently, but rather in bursts. That is, the opening of one airway may initiate the opening of several more peripheral airways. When the air pressure implied on the lung (P_B) exceeds the opening threshold of an airway, all the daughter airways that had a smaller threshold would be opened immediately (one can neglect the time of opening an airway in this situation). The number of airways, and hence the size of the recruited alveolar volumes involved in such an opening sequence, depends on the size of subtrees, and can vary substantially. The sequential activation of alveoli triggered by overcoming a

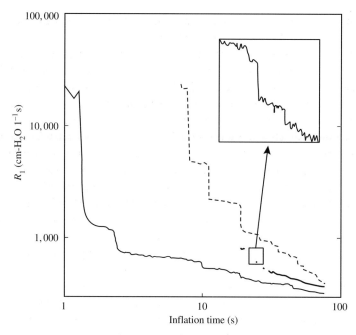

Fig. 3.4 Terminal airway resistance as a function of inflation time (from [9]). Reprinted by permission from Nature, copyright 1994 Macmillan Magazines Ltd.

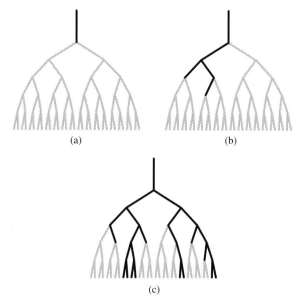

Fig. 3.5 Typical development of avalanches in the lung model. (a) After the main alveola has opened, a small part of the lung also opens (b) due to lower critical pressure in that region. (c) The situation after a large avalanche: about the half of alveoli are open (after [9]). Reprinted by permission from Nature, copyright 1994 Macmillan Magazines Ltd.

local threshold can be considered as an avalanche. The existence of avalanches often leads to power-law dependences (i.e., criticality).

To test the above interpretation of the lung-opening phenomena a simple SOC model, similar to one considered for the evolution, can be constructed. Consider the branched lung structure, as shown in Fig. 3.5(a). Each airway is assigned an opening threshold P_i which takes a random value from the interval $[0, 1]$. Then we set $P_B = 1$ and lower it until it reaches the opening threshold ($P_B = P_0$) of the main alveola. At this point the main alveola opens (Fig. 3.5(a)). Next, the two daughter airways are checked: they are opened if their threshold is lower that the actual bronchial pressure (P_B). This process is repeated until there is no more airway to open (Fig. 3.5(b)), i.e., all airways which are in contact with the opened part of the lung have $P_i > P_B$. At this point we consider this opening event (the avalanche) finished. Then P_B is incremented and the process is iterated so that larger and larger part of the lung opens. The newly opened alveolar volume in an avalanche can be significant with respect to the total lung volume, as is demonstrated in Fig. 3.5(c). The statistical analysis of the above model yields results which are consistent with the ones obtained experimentally. One can conclude that the process of lung inflation can be well treated as self-organized critical.

The examples reviewed in this section show that the general framework of self-organized critical phenomena proves useful for describing a range of dynamical biological processes. In many situations simple SOC models containing some sort of threshold mechanisms can be easily constructed to give a theoretical interpretation of the observed phenomena.

References

[1] P. Bak, C. Tang, and K. Wiesenfeld. Self-Organized criticality: an explanation of $1/f$ noise. *Phys. Rev. Lett.* 59: 381, 1987.

[2] S. R. Nagel. Instabilities in a sandpile. *Rev. Mod. Phys.* 64: 321–325, 1992.

[3] D. Raup. *Extinction: Bad Genes or Bad Luck?* W. W. Norton, London, 1992.

[4] P. Bak and K. Sneppen. Punctuated equilibrium and criticality in a simple model of evolution. *Phys. Rev. Lett.* 71: 4083–4086, 1993.

[5] K. Sneppen, P. Bak, and M. J. Jensen. Evolution of a self-organized critical phenomenon. *Proc. Natl. Acad. Sci.* 92: 5209, 1995.

[6] M. Marsili, P. De Los Rios, and S. Maslov. Expansion around the mean field solution of the bak-sneppen model. *Phys. Rev. Lett.* 80: 1457, 1998.

[7] H. Flyvbjerg, K. Sneppen, and P. Bak. Mean field theory for a simple model of evolution. *Phys. Rev. Lett.* 71: 4087–4090, 1993.

[8] M. F. Shlesinger and B. J. West. Complex fractal dimension of the bronchial tree. *Phys. Rev. Lett.* 67: 2106–2108, 1991.

[9] B. Suki, A.-L. Barabási, Z. Hantos, F. Peták, and H. E. Stanley. Avalanches and power-law behaviour in lung inflation. *Nature* 368: 615–618, 1994.

4 Patterns and correlations

4.1 Bacterial colonies

A. Czirók

4.1.1 Introduction

Bacteria are often said to be typical unicellular organisms and thought to live independently from each other. However, in their natural living places like on soil or rock surfaces bacteria form colonies and, as we will see, these colonies alter their form depending on the environmental conditions. This response cannot be understood by assuming that individual cells simply stay close to each other as the consequence of being difficult to move. In fact, some of the complex biochemical processes performed by bacteria could not be possible without forming organized groups [57–59]. In some cases bacteria differentiate into specialized cell types (Fig. 4.1) within the colonies resulting in increased resistance to antibacterial agents or ability of invading eukaryotic hosts.

In addition to these interesting microbiological aspects, the study of bacterial colony formation can give certain insights into the formation of self-organized biological structures in general. When trying to understand the function and operation of complex interacting networks like the nervous system or the intracellular signal transduction pathways, we often lack the appropriate experimental methods and abstract

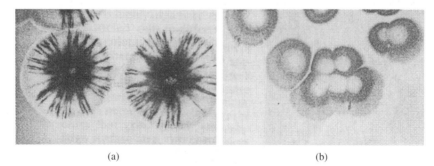

(a)   (b)

Fig. 4.1 Examples of inhomogeneous protein expression in bacterial colonies grown on agar surface. (a) Some cells of *Chromobacterium violaceum* produce the pigment violacein. (b) These *Escherichia coli* colonies carry genetically engineered DNA sequences encoding the enzyme β-galactosidase; the colony turns blue where the enzyme is expressed (after [57]).

formalism. In the case of colonies, however, we can hope that the interactions are still simple enough to be captured by quantitative models and that the collective behaviour can be explored with the available computational power.

A basic problem of such studies is related to the very different length scales involved in the problem: The self-organized structures are macroscopic, consisting up to 10^{10} bacteria with a typical length scale of a few cm. However, these patterns are determined by the bacterial behaviour and interactions at the microscopic scale of the individual cells. Given this large difference in the length scales, it is usually not straightforward to predict the effect of a certain microbiological feature on the macroscopic behaviour of the colony. In fact, as we will show, many of the microscopic details of the underlying system can turn out to be irrelevant to the dynamics of the large-scale structures. This *universality* of the pattern formation process explains the fact that the same patterns may develop in completely different systems.

As a case study, within this chapter we would like to describe and understand the colony formation of a few bacterial strains displaying somewhat similar macroscopic behaviour. First we overview some of the microbiological background (after [52]) and then discuss the relevant experimental observations. In the remaining parts we investigate one by one the development of the various morphologies and the corresponding models.

4.1.2 Bacteria in colonies

In the simplest experimental studies of colony formation, bacteria are grown on the surface of agar gels, as Fig. 4.2 shows. If the surface is dry and the cells cannot move, then the time needed to spread over the substrate can take several weeks. Since this time period is long compared to the duplication time of the cells, the *proliferation* of the bacteria is the key factor determining the colony morphology. In the other extreme case, i.e., if the gel is soft or the bacteria produce surfactants, the colony spreads over the substrate in a few hours and the bacterial *motion* and its *chemotactic* modulation are important. In the general case, both of these effects must be considered to understand the response of the colonies to changes in their environments.

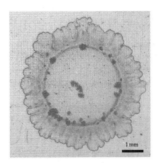

Fig. 4.2 A growing bacterial colony (*Bacillus subtilis 168*) on the surface of a dry agar gel. Cells were inoculated in the central circular region of the picture (after [16]).

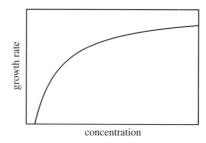

Fig. 4.3 Schematic representation of the bacterial growth rate vs the nutrient concentration in the environment.

Microbiological background

Proliferation   As revealed by experiments in fermentors where microorganisms are grown in a steady, homogeneous liquid environment, the increase of the number and total mass of bacteria strongly depends on the nutrient concentration. In fact, the *rate* of growth (number of cell divisions within a population of unit size during a unit time interval) increases with the nutrient concentration in a hyperbolic manner reminiscent to Michaelis–Menten kinetics (Fig. 4.3). This finding probably indicates that the nutrient uptake is mediated by transport proteins. There is, however, a minimal concentration of nutrients required for bacterial proliferation, since not *all* the uptaken free energy is converted to growth as there is a finite amount required to maintain the intracellular biochemical processes [36, 50].

Motility   Procaryotes move in aquatic environments by rotating their flagella, which are rather rigid membrane-bound helical protein polymers. Bacteria can have either only one flagellum (monotrichous), a pair of flagella at the opposite cell poles (amphitrichous), clusters of flagella at the poles (lophotrichous), or flagella uniformly distributed over the cell membrane (peritrichous).

Detailed microscopic investigations revealed that the direction of flagellar rotation determines the nature of bacterial movement. A single, polar flagellum rotates counterclockwise (when viewed from outside the cell) during normal forward movement, while the cell body rotates in the opposite direction. In general, multiflagellated cells operate in a similar fashion: each flagellum rotates counterclockwise, and they form a bundle that propels them forward. This forward motion is interrupted by short intervals of '*tumbling*' when most flagella rotate in the opposite direction disrupting the bundles, which results in a random change in the orientation of the cell.

A very different type of motility, so-called *gliding*, is employed by many strains when moving on surfaces. Surprisingly little is known about this type of motion as there are no visible cellular structures associated with it. In fact, gliding motility varies greatly in the nature of the motion, which may indicate that more than one mechanism for gliding motility exists. For example, bacteria such as *Myxococcus* and *Flexibacter* glide along in a direction parallel to the long axis of the cells. Others display a screw-like motion (e.g., *Saprospira*) or even move in a direction perpendicular to the long axis (*Simonsiella*). Gliding bacteria secrete viscous extracellular glycoconjugates (slime) but those do not appear to propel them directly; rather, they probably attach them to the substrate and lubricate the surface for more efficient movement. There is some evidence that differences in surface tension can propel *Myxococcus xanthus*, which may secrete a surfactant at its end opposite to the direction of the movement.

Chemotaxis. Bacteria rarely move aimlessly since they are attracted by nutrients like sugars and amino acids and are repelled by many harmful substances and metabolic waste products. They can also respond to other environmental factors like temperature, light or oxygen concentration. This movement toward chemical attractants or away from repellents is known as *chemotaxis*.

The chemotactic response is a stochastic process, since bacteria exposed to chemical gradients modulate their tumbling frequency. Tumbling is repressed when moving towards higher concentration of chemoattractants, i.e., they continue to move forward for a more extended period of time. In fact, the swimming behaviour is influenced by changes of the chemical concentration *in time*: the molecular machinery in a bacterium compares its current environment with that experienced a few moments previously. The underlying biochemical processes have been identified in detail for *Escherichia coli*.

Morphology diagram

The number of environmental parameters influencing colony development (temperature, humidity, chemical composition of the substrate, etc.) is rather large, and various sets of parameters can give rise to different morphologies even for the very same bacterial strain. In systematic experiments aimed at exploring the effect of environmental factors on colony behaviour, the concentration of the *agar* and *nutrients* were changed systematically in the substrate [16, 29, 51]. The former component is related to the consistency of the gel and thus to the motility of the bacteria and the diffusibility of the nutrients. As we have seen, the nutrient concentration determines the proliferation rate of the cells (Fig. 4.3). As a result of these experiments, *morphology diagrams* were constructed in which a characteristic colony shape was assigned for each pair of the parameters.

The first such experiments have been performed with a wild type *Bacillus subtilis* strain [29, 51] (later denoted as OG-01 [46], Fig. 4.4(a)), a non motile mutant of the same strain (OG-01b [51], Fig. 4.4b) and a strain[1] [16] denoted after its characteristic morphology as *Paenibacillus dendritiformis* (Fig. 4.5). The morphology diagrams of these three strains show certain similarities. The growth is compact at high concentrations of the nutrients and becomes branching under nutrient deficient conditions. However, the diagrams do differ in many important aspects.

---

[1] This strain was first thought to be a variant ('$\mathcal{T}$ morphotype') of *B. subtilis* 168 [14], but recently it has been genetically identified as a new strain [61] coexisting in the original stock with *B. subtilis* [55].

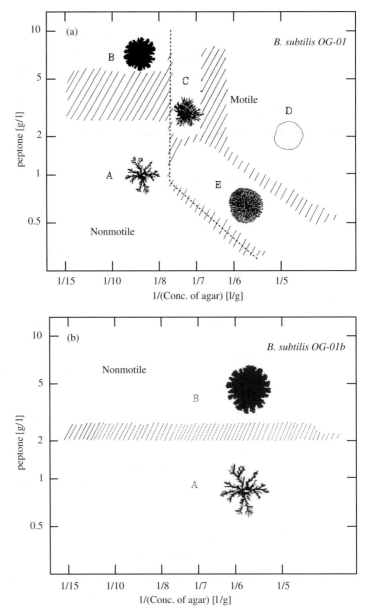

Fig. 4.4 Morphology diagram of the strain *Bacillus subtilis* OG-01 (a) and a non-motile mutant OG-01b (b) as a function of peptone and agar concentration. The dashed line in (a) indicates the boundary of the active movement of bacterial cells inside the colonies. The morphologies are classified as follows: fractal (A), compact with rough boundary (B), branching with periodic growth phases (C), compact with diffuse boundary (D) and dense branching (E). In the case of the non-motile strain the regions A and B seen in (a) expand laterally, while regions C, D and E disappear (after [51]).

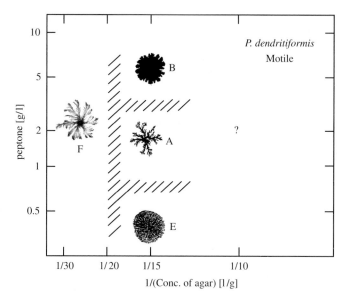

Fig. 4.5 Morphology diagram of *Paenibacillus dendritiformis* (*T* morphotype) colonies as a function of peptone and agar concentration, drawn using data published in [16]. On hard substrate a new, twisted morphology appears (F). On soft agar even the *phenotype* of the bacteria changes as discussed in §4.1.5.

P. dendritiformis colonies exhibit dense branching morphology (*DBM*) characterized by thin, radial branches below 1 g/l peptone concentration. When grown on hard agar (above 20 g/l), the branches are *curved*. In contrast, *B. subtilis* OG-01 cells are non-motile on hard agar. On softer agar, the colony can display periodic growth patterns and *DBM*, but not at low nutrient concentrations. These differences demonstrate that certain microbiological features do have a strong influence on the colony morphology and, hence, it is not determined entirely by the environmental conditions. At the same time, the pure existence of such diagrams and the reproducibility of the patterns suggests that the behaviour of the colonies is still determined by quantitative laws.

Statistical measures of morphology

To distinguish the various stochastic structures and classify their morphology appropriate statistical measures are needed. One widely applied measure is the *density–density correlation* function, defined as the conditional probability

$$C_\varrho(r) \equiv \frac{\langle \varrho(r\boldsymbol{e} + \boldsymbol{a})\varrho(\boldsymbol{a})\rangle_{e,a}}{\langle \varrho(\boldsymbol{a})\rangle_a}, \qquad (4.1)$$

where *e* is an arbitrary unit vector and $\varrho(\boldsymbol{a})$ denotes the 'density' of the object at the point *a*: ϱ is 1 if *a* belongs to the object and zero otherwise. We use the notation $\langle f(\boldsymbol{a})\rangle_a$ for the average of any quantity f over all possible values of *a*. Then, by definition, $C_\varrho(0) = 1$, and if the object has a finite size L, then $C_\varrho(r > L) = 0$.

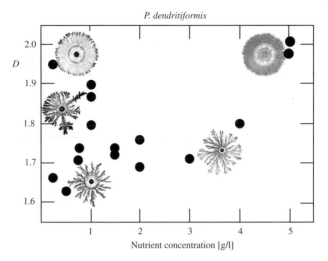

Fig. 4.6 Quantitative characterization of the morphology of *P. dendritiformis* colonies grown on substrates containing 20 g/l agar and various amount of nutrients. The filled circles show the fractal dimension (D) of the colonies vs the initial nutrient concentration. Representative colony images are placed in the vicinity of the position corresponding to their fractal dimension and growth condition. On nutrient-rich substrates the colonies are compact ($D \approx 2$), while-lowering the nutrient concentration fractal branches develop ($D \approx 1.7$). Below 1 g/l peptone concentration the fractal patterns coexist with the dense branching morphology characterized by thin radial branches and $D \approx 2$. This coexistence of the morphologies can sometimes be observed within the same colony, as the leftmost inset demonstrates. The formation of these remarkable patterns is discussed in §4.1.4.

In many cases C_ϱ decays as a power law

$$C_\varrho(r) \sim r^{-\alpha} \tag{4.2}$$

for $1 \ll r \ll L$. Within these limits such objects are *scale-invariant* (fractal), since the morphology, i.e., the correlation function eqn (4.2), is invariant under the $r' = br$ rescaling as

$$C'_\varrho(r') = C_\varrho(br) \sim b^{-\alpha} r^{-\alpha} \sim C_\varrho(r), \tag{4.3}$$

where C'_ϱ denotes the correlation function of the rescaled object. As described in §2.1, the fractal dimension D is related to the exponent α as $D = 2 - \alpha$ [65]. The calculated fractal dimensions of *P. dendritiformis* colonies are shown in Fig. 4.6. Since both the compact and *DBM* morphologies are characterized by $D \approx 2$, in some cases further measures complementing C_ϱ are needed.

To characterize the irregular surface of compact objects (Fig 4.7), the *height–height correlation* function can be used, defined as

$$C_h^2(x) \equiv \left\langle [h(x+a) - h(a)]^2 \right\rangle_a, \tag{4.4}$$

where the surface is denoted by $h(x)$, and it is measured as a distance from a certain line or point of reference. Again, the quantity $C_h(x)$ has a clear meaning: it is the

Fig. 4.7 Typical rough colony boundaries of various bacterial strains. (a) *Bacillus subtilis* 168, after [64], and (b) *Escherichia coli* , after [63]. Copyright 1990 reprinted with permission from Elsevier Science, grown on a solid substrate rich in nutrients.

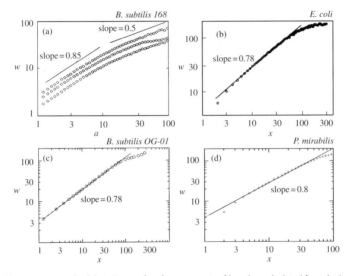

Fig. 4.8 The average standard deviation w of surface segments of length x calculated for colonies of (a) *B. subtilis* 168 (after [64]), (b) *Escherichia coli* (after [63]), Copyright 1990 reprinted with permission from Elsevier Science. (c) *Bacillus subtilis* OG-01 (after [67]) and (d) *Proteus mirabilis* (calculated by the author from images published in [54]). The linear regime on the double-logarithmic plots indicates the scaling behaviour eqn (4.6). The plotted values are given in arbitrary units (pixels).

expected change in h over a distance x. Another possibility is to calculate the *average standard deviation* w of the surface in a window of linear size x, i.e.,

$$w^2(x) \equiv \left\langle \langle h^2(a)\rangle_{u\leq a\leq u+x} - \langle h(a)\rangle^2_{u\leq a\leq u+x} \right\rangle_u. \tag{4.5}$$

Figure 4.8 shows some typical examples of the correlation functions calculated for the boundary of bacterial colonies. In many cases, irrespective of the bacterial strain,

$$w(x) \sim C_h(x) \sim x^H \tag{4.6}$$

was found with

$$H = 0.8 \pm 0.05 \tag{4.7}$$

within quite a broad interval of x, i.e., the interface is a *nontrivial self-affine* curve with long-range correlations decaying as a power law. The insensitivity of H to the underlying biological system gives the major support for the following models predicting 'universal' behaviour. As was mentioned in the introduction, in our context *universality* means that the large-scale patterns are self-organized, and not influenced by many details of the underlying systems.

4.1.3 Compact morphology

As was shown in the previous section, if the nutrients are abundant in the substrate, then *compact* colonies with either smooth or irregular perimeters are formed. In this section we investigate this simplest case of colony morphogenesis in detail. The results will serve as a starting point for the studies of more sophisticated colonies where such detailed mathematical analysis cannot be carried out.

Spreading of bacteria: the Fisher-Kolmogorov equation

In the simplest example of bacterial spreading, nutrients are abundant and the substrate is wet enough to allow bacterial motion. Under such circumstances the cells proliferate and migrate across the dish within a few hours (see, for example, region D in Fig. 4.4(a)). Videomicroscope tracking of the bacteria revealed that their trajectories are *random walks* [68]. Therefore, the interaction between the cells must be negligible and the time dependence of the bacterial density ϱ (number of cells in a unit surface are) can be described by the Fisher-Kolmogorov equation [49]

$$\partial_t \varrho = D_\varrho \nabla^2 \varrho + f(\varrho). \tag{4.8}$$

The first term on the right-hand side of eqn (4.8) describes the random *translocation* of bacteria. The diffusion coefficient D_ϱ is related to the microscopically measurable mean-squared displacement $d^2(t)$ of the individual cells during a time period t as

$$\overline{d^2(t)} = 2D_\varrho t, \tag{4.9}$$

where the overline represents ensemble average among cells.

The bacterial *multiplication* is incorporated in f. When the bacterial density is small, the cells proliferate with a fixed rate resulting in an exponential growth. Thus, $f(\varrho) = r(c)\varrho$ for small ϱ, where r depends on the local nutrient concentration c. As we discussed in §4.1.2, $r \sim c$ holds for small values of c. In practice, even with unlimited nutrient supply, the cell density cannot increase above a certain threshold ϱ_*, e.g., because of the accumulation of toxic metabolites. Throughout this section we measure the cell density in units such that $\varrho_* = 1$. Thus, the growth rate must decrease as $\varrho \to 1$, and hence $f(1) = 0$. The particular form of f will be unimportant, but where it is needed, we use the logistic function $f(\varrho) = r\varrho(1 - \varrho)$ as an example. Although eqn (4.8) is very simple, we discuss it here because it serves as a prototype of such problems, and it is still far from being trivial even in one dimension.

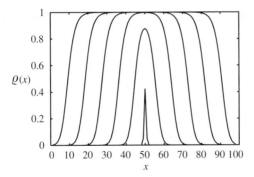

Fig. 4.9 Typical result of the numerical integration of the Fisher eqn (4.8) starting from a localized perturbation ($D_\varrho = 1$, $f(x) = x - x^2$, i.e., $r = 1$). The $\varrho(x,t)$ curves are plotted for $t = 0, 5, 10, 15, 20$ and 25. The domain grows with a stationary speed of $v = 2$.

Numerical integration of eqn (4.8) starting from a localized (e.g., Gaussian) perturbation results in a growing domain expanding with a constant speed $v \approx v_*$, where

$$v_* = 2\sqrt{rD_\varrho} \tag{4.10}$$

(see Fig. 4.9). When, in order to calculate v, the expanding domain of $\varrho(x,t) \sim 1$ is written in a moving frame of reference as $\tilde{\varrho}(u,t) = \varrho(x,t)$, where $u = x - vt$ and without loss of generality $v > 0$, $\varrho(-\infty) = 1$ and $\varrho(\infty) = 0$ is assumed, one obtains

$$\partial_t \tilde{\varrho} = D_\varrho \tilde{\varrho}'' + v\tilde{\varrho}' + f(\tilde{\varrho}), \tag{4.11}$$

where the prime denotes differentiation with respect to u.

Surprisingly, eqn (4.11) has a *stationary* solution for *any* value of $v \geq v_*$, i.e., v cannot be derived by the stationarity condition imposed on $\tilde{\varrho}$. This feature constitutes the 'velocity selection problem' associated with similar reaction–diffusion equations. In the case of the Fisher equation the dynamics selecting the front propagating with $c = c_*$ from the set of possible stationary solutions of eqn (4.11) was published in [7] and discussed in detail in [9].

Experimental results The velocity selection in real bacterial spreading has been experimentally investigated with *B. subtilis* grown in region E of the morphology diagram (Fig. 4.4(a)). The spreading speed of the colony was measured for various nutrient concentrations [68], which determines the growth rate as $r \sim c$. The bacterial diffusion parameter D_ϱ was calculated using eqn (4.9) and was found to be fairly constant while changing c within an order of magnitude. Thus, eqn (4.10) predicts

$$v \sim \sqrt{c}. \tag{4.12}$$

As described in [68], colonies started growing about eight hours after the inoculation. As the colony expanded, its contour became circular, its growth speed increased and asymptotically approached a finite value. This asymptotic growth speed was measured when the diameter of the colony reached 50 mm. The obtained experimental

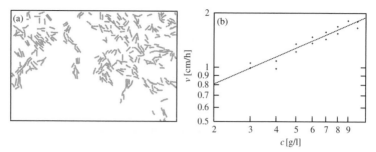

Fig. 4.10 (a) Microscope image of the fuzzy interface of a *B. subtilis* colony. The scale of the figure is 0.2 mm in width. (b) The growth speed v of the interface vs the nutrient concentration. The dashed line shows the expected square root-like behaviour (after [68]).

Fig. 4.11 Long bundles of chains of individual cells constituting the colony of the *B. subtilis* OG-01 strain grown on hard agar. Note the abrupt change in cell density at the boundary (after [67]).

data are shown in Fig. 4.10, and within the limited, experimentally accessible range are in agreement with eqn (4.12).

Cell–cell interactions In most interesting cases, however, individual bacteria are not independent during the spreading of the colony. For example, when the cells are not motile, then there is an abrupt change in the cell density at the border of the colony, as Fig. 4.11 shows. In this case the propagation of the boundary is driven mainly by the expansion of the cell volumes *inside the colony*: If the bacteria cannot expand to their preferred size, then they exert mechanical pressure on their surrounding as discussed in [36]. Thus, for large cell densities this pressure p is proportional to $\varrho - \varrho_0$, where ϱ_0 is the threshold density determined by the close-packed configuration of the cells on the substrate. Since the attachment of the bacteria to the substrate cannot withstand arbitrarily large stresses, at least for large p, the displacement of the cells will be proportional to ∇p as described by

$$v = D_0 \nabla (\varrho - \varrho_0). \tag{4.13}$$

The D_0 prefactor plays a similar role in the following as D_ϱ in eqn (4.8). Hence, in this case the time development of the bacterial density is described by a modified

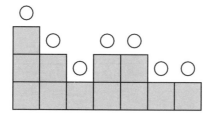

Fig. 4.12 Generation of the Eden cluster. In each step of the process one of the lattice sites (marked by circles) next to the 'colony' (grey) is selected with uniform probability and filled.

form of the Fisher–Kolmogorov equation

$$\partial_t \varrho = \begin{cases} D_0 \nabla^2 \varrho + f(\varrho) & \text{for } \varrho > \varrho_0 \\ f(\varrho) & \text{otherwise.} \end{cases} \tag{4.14}$$

Again, in one dimension eqn (4.14) gives travelling fronts with a constant velocity v. In the moving frame of reference $u = x - vt$ for the stationary solution of eqn (4.14) and we obtain

$$D_0 \tilde{\varrho}'' + v \tilde{\varrho}' + f(\tilde{\varrho}) = 0 \tag{4.15}$$

inside the 'colony', i.e., for $u < 0$. At $u = 0$ the boundary conditions are

$$\tilde{\varrho}(0) = \tilde{\varrho}_0 \tag{4.16}$$

$$v \tilde{\varrho}_0 = D_0 \tilde{\varrho}'(0), \tag{4.17}$$

which ensures the conservation of $\tilde{\varrho}$. In such cases, however, as Figs. 4.7 and 4.11 demonstrate, the colony boundary does not remain smooth, as the growth is strongly influenced by the various inhomogeneities of the environment.

Self-affine interface roughening

The experimental finding that the boundary of bacterial colonies shows self-affine properties was not completely unexpected, since one of the earliest methods of generating self-affine objects had related biological motivation. In the variants of the Eden model – published in 1961 [25] – an assembly of 'cells' is grown on a lattice with the following rules (see Fig. 4.12). In each step of the process one of the lattice sites next to the 'populated' areas is chosen randomly and occupied.[1] Although this stochastic dynamics is very simple, the resulting objects are quite irregular, as Fig. 4.13 shows. Large-scale numerical studies revealed that for the surface of the two-dimensional Eden 'colony' eqn (4.6) holds with $H = 1/2$ [33].

The Kardar–Parisi–Zhang (KPZ) equation To understand this frequent occurrence of self-affine curves in surface- roughening phenomena, an approach based on

---

[1] In fact, the model has slightly different versions assigning different probabilities for occupying the perimeter sites depending on the local configuration of the colony. However, the statistical features of the developing clusters are the same for all variants of the model in the asymptotic limit [33].

Fig. 4.13 A typical colony in the Eden model grown on a strip of 256 lattice units (after [63]). Copyright 1990 reprinted with permission from Elsevier Science.

stochastic differential equations can be applied. It turns out to be fruitful to simplify the problem defined by eqn (4.14) and focus our attention upon the interface itself. The motion of interfaces in the presence of external fluctuations or in an inhomogeneous environment can be described [34] by

$$\partial_t h = \nu \partial_x^2 h + \frac{\lambda}{2}(\partial_x h)^2 + \nu + \eta, \tag{4.18}$$

where the terms at the right-hand side of eqn (4.18) have the following physical meaning.

The surface tension term, $\nu \partial_x^2 h$, tends to smoothen the interface and does not permit discontinuities (large jumps) in h.

The nonlinear $\lambda(\partial_x h)^2 + \nu$ terms reflects the *isotropy* of the growth. If the displacement is *perpendicular* to the interface, then using the notations of Fig. 4.14,

$$\Delta h \approx \nu \Delta t \sqrt{1 + tg^2 \phi} \approx \nu \Delta t + \frac{\nu \Delta t}{2}(\partial_x h)^2 \tag{4.19}$$

holds for small slopes.

Finally, $\eta(x, t)$ represents some sort of *noise*. Various type of models can be distinguished based on the assumed properties of this term: the simplest choice is a noise *uncorrelated* both spatially and temporally defined as

$$C_\eta(x, t) \equiv \langle \eta(x', t')\eta(x + x', t + t')\rangle_{x',t'} = 2D\delta(x)\delta(t), \tag{4.20}$$

where the average $\langle \ldots \rangle_{x',t'}$ is calculated over all possible values of x' and t', and the $\delta(x)$ measure satisfies $\int_{-\infty}^{\infty} \delta(x) = 1$ and $\delta(x) = 0$ for all $x \neq 0$. The infinite standard deviation ($C_\eta(0, 0) \equiv \langle \eta^2 \rangle$) is only a formal consequence of the continuum limit[1]. In the case of the Eden model the noise comes from the random choice of growth segments, which is an uncorrelated process by definition.

---

[1] If x and t are discrete variables, then eqn (4.20) can be written in the more obvious form of $C_\eta(x, t) = \delta_{x0}\delta_{t0}$, where the Kronecker symbol $\delta_{ij} = 1$ if $i = j$ and zero otherwise. The continuum limit eqn (4.20) can be derived from the requirement that the standard deviation of the stationary noise averaged over a finite interval must be finite and proportional to the length of the interval, i.e.,

$$\left\langle \left(\int_x^{x+\ell} \eta(a)\, da \right)^2 \right\rangle_x = \left\langle \int_0^\ell da \int_0^\ell db\, \eta(x+a)\eta(x+b) \right\rangle_x$$

$$= \int_0^\ell da \int_0^\ell db\, \langle \eta(x+a)\eta(x+b)\rangle_x = \int_0^\ell da \int_0^\ell db\, C_\eta(a-b) \sim \ell. \tag{4.21}$$

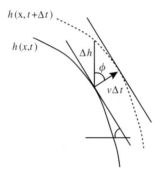

Fig. 4.14

There are quite a few analytical methods available to solve eqn (4.18). Probably the simplest way is based on the calculation of the probability distribution of the steady-state height fluctuations [30]. If the stochastic variables $y(t) = \{vy_i(t)\}, i = 1, \ldots, n$ are governed by the Langevin equation

$$\frac{dy}{dt} = A(y) + \eta(t) \tag{4.22}$$

where A is any – usually nonlinear – vector function and $\eta(t)$ is an uncorrelated noise

$$C_{ij}(t) \equiv \langle \eta_i(t')\eta_j(t+t')\rangle_{t'} = 2D\delta(t)\delta_{ij} \tag{4.23}$$

in analogy with eqn (4.20), then the time evolution of the probability distribution $P(y, t)$ is given by the Fokker–Planck equation

$$\partial_t P(y, t) = -\text{div}[A(y)P(y, t) - D(\nabla P)(y, t)] \tag{4.24}$$

where $\text{div } A = \sum_{i=1}^{n} \partial_{y_i} A_i$ and $(\nabla P)_i = \partial_{y_i} P$.

Equation (4.24) can be generalized to describe the time evolution of the probability density of a continuous $h(x, t)$ function instead of a finite number of variables represented by $y(t)$. The gradient, which assigns a vector field to a scalar field, can be replaced by the functional derivative $\mathcal{D}$ which assigns a function to an $I(f)$ functional as

$$I(f + \epsilon) = I(f) + \int (\mathcal{D}I)(x)\epsilon(x)\,dx \tag{4.25}$$

for arbitrary $\epsilon(x)$, in the $\epsilon(x) \to 0$ limit. In a similar fashion, the derivative $\mathcal{D}\mathcal{F}$ of an operator $\mathcal{F}$ is defined through

$$\lim_{\epsilon(x)\to 0} [\mathcal{F}(f + \epsilon)](x) = [\mathcal{F}(f)](x) + \int [(\mathcal{D}\mathcal{F})(f)](x, y)\epsilon(y)\,dy, \tag{4.26}$$

since for a fixed x the $[\mathcal{F}(f)](x)$ quantity is a functional of f. Finally, the generalization of the div operation is

$$(\text{div }\mathcal{F})(f) = \lim_{\delta\to 0} \int dx \int_{x-\delta}^{x+\delta} dy\, [(\mathcal{D}\mathcal{F})(f)](x, y). \tag{4.27}$$

With these notations eqn (4.24) can be written in the form

$$\partial_t P(h, t) = -\mathrm{div}\left[\mathcal{F}(h)P(h) - D(\mathcal{D}P)(h)\right],\tag{4.28}$$

where $\mathcal{F}(h) = \partial_x^2 h + \frac{\lambda}{2}(\partial_x h)^2$.

In general, it is very hard to solve equations like eqn (4.28). However, in our particular case one can show that

$$P(h) = \exp\left(-\frac{1}{2}\int(\partial_x h)^2\right)\tag{4.29}$$

is a stationary solution, i.e., in the stationary regime of eqn (4.18) the probability of finding a particular $h(x)$ curve is proportional to $P(h)$ as given in eqn (4.29). To see this let us substitute eqn (4.29) into eqn (4.28). In the $\epsilon \to 0$ limit

$$P(h + \epsilon) = \exp\left[-\frac{1}{2}\int(\partial_x h)^2 - \int(\partial_x h)(\partial_x \epsilon)\right] = P(h) - P(h)\int(\partial_x h)(\partial_x \epsilon)\tag{4.30}$$

holds. Performing a partial integration of the last term of eqn (4.30) and using definition eqn (4.25)

$$(\mathcal{D}P)(h) = P(h)\partial_x^2 h\tag{4.31}$$

is obtained. Thus the second term on the right-hand side of eqn (4.28) cancels the λ-independent contribution of the first term. The remaining, λ-dependent expression is

$$-\frac{\lambda}{2}\mathrm{div}\left[(\partial_x h)^2 P(h)\right] = -\frac{\lambda}{2}\int(\partial_x h)^2 \mathcal{D}P(h) - \frac{\lambda}{2}P(h)\,\mathrm{div}(\partial_x h)^2.\tag{4.32}$$

By substituting eqn (4.31) into eqn (4.32) and performing a partial integration one can show that the first term on the right-hand side of eqn (4.32) is identically zero. A similar statement holds for the remaining term as well. Based on eqn (4.26),

$$\left[\mathcal{D}(\partial_x h)^2\right](x, y) = 2(\partial_x h)(x)\delta'(x - y).\tag{4.33}$$

Therefore,

$$\begin{aligned}
\mathrm{div}(\partial_x h)^2 &= 2\int dx \int dy\ (\partial_x h)(x)\delta'(x - y)\\
&= -2\int dx \int dy\ (\partial_x^2 h)(x)\delta(x - y)\\
&= -2\int dx\ (\partial_x^2 h)(x) = 0.
\end{aligned}$$

The probability distribution eqn (4.29) tells us that the consecutive slopes of h are independent variables, therefore in the $t \to \infty$ limit $h(x)$ is a random walk characterized by $H = \frac{1}{2}$. Moreover, the solution eqn (4.29) remains the same if the $(\partial_x h)^2$ term in eqn (4.18) is replaced with any linear combination of nonlinear terms of the type $(\partial_x h)^{2k}$ with a positive integer k. This robust analytical result for surface roughening driven by temporally uncorrelated fluctuations illustrates best what we mean under universality and explains why the various versions of the Eden model with slightly different rules give the same asymptotic behaviour. Unfortunately, the $H = 1/2$ result is not compatible with the experimental finding eqn (4.7).

Quenched noise   For bacterial colonies – and for many other physical systems as well – the assumption of uncorrelated noise is not realistic. If the major factor modulating the interface velocity is the local inhomogeneity of the agarose gel, then this might be modelled as uncorrelated in space, but it is certainly *correlated in time*. In fact, η is then a function of the position of the interface

$$\eta(x, t) = 2D\tilde{\eta}(x, h(x, t)) \tag{4.34}$$

with

$$C_{\tilde{\eta}}(x, y) = \Delta(x)\Delta(y), \tag{4.35}$$

where $\Delta(u) \sim 1$ holds for some finite interval of u around zero.

If such a 'quenched' noise is incorporated into eqn (4.18), then the behaviour of the equation is significantly changed. By appropriate choice of the time and length units and using eqn (4.19) all the parameters λ, ν and v can be transformed out, i.e., it is sufficient to investigate the $\lambda = \nu = v = 1$ case

$$\partial_t h = \partial_x^2 h + \sqrt{1 + (\partial_x h)^2} + \eta, \tag{4.36}$$

where the full nonlinear term representing the lateral growth is included. The only remaining parameter of the problem is D in eqn (4.34), which determines the average magnitude of the noise as $\sqrt{C_{\tilde{\eta}}(0, 0)} = \sqrt{2D}$. Now, two extreme cases can be distinguished.

If $D > D_* \sim 1$ then the interface encounters points where it becomes pinned and it will experience that particular condition for a long time until it is released by the additional 'pulling' force caused by the advancement of the neighbouring segments of the interface. If the density of such pinning sites is high enough, then eventually the propagation of the whole surface can be blocked. The shape of the frozen interface is then determined by the distribution of the pinning sites in the substrate and becomes independent of the dynamics of the growth. In fact, the surface roughening can be mapped onto a specific percolation problem (e.g., see [8]). Since the interface is blocked at each possible segment, it is adjacent to a cluster of pinning sites. Due to the single-valuedness of h, the cluster is *directed*: it spans in a certain direction, turns up and down, but never back.[1]

Such a *directed percolation cluster* is characterized by two correlation lengths, being parallel ($\xi_\parallel$) and perpendicular ($\xi_\perp$) to the interface (Figs. 4.15, 4.16). As

---

[1]   The description using single-valued functions can be valid only for processes resulting in rather flat surfaces, i.e., for $H < 1$. Otherwise 'overhangs' can be generated which are ignored here.

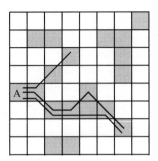

Fig. 4.15 Examples of directed percolation clusters leaving site A. Such clusters consist of filled neighbour sites and extend from left to right.

(a)   (b)

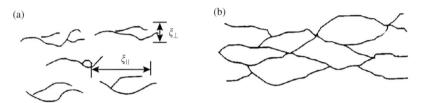

Fig. 4.16 Directed percolation clusters below (a) and above (b) the critical density of the filled lattice sites at which a cluster spanning the entire system appears. Below this percolation threshold the directed percolation clusters have a typical length $\xi_\parallel$ (parallel to the distinguished direction) and a typical width $\xi_\perp$.

exhaustive studies [26] revealed, these correlation lengths diverge in the vicinity of the threshold density of the pinning sites p_c as

$$\xi_\parallel \sim |p - p_c|^{-\nu_\parallel}, \qquad \xi_\perp \sim |p - p_c|^{-\nu_\perp} \tag{4.37}$$

with

$$\nu_\parallel = 1.733 \quad \text{and} \quad \nu_\perp = 1.097. \tag{4.38}$$

Complete blocking of the interface propagation appears when $\xi_\parallel$ becomes equal to the system size L. The width w of such a path is in the order of $\xi_\perp$, hence

$$L^H \sim w \approx \xi_\perp \sim |p - p_c|^{-\nu_\perp} \sim \xi_\parallel^{\nu_\perp/\nu_\parallel} \approx L^{\nu_\perp/\nu_\parallel}, \tag{4.39}$$

i.e., for the blocked interface one obtains

$$H = \frac{\nu_\perp}{\nu_\parallel} = 0.633 \tag{4.40}$$

in accord with the numerical simulation of eqn (4.36) with $D \gg D_*$ [23].

In the $D \ll D_* \sim 1$ case the interface is never pinned and advances with a steady velocity. Since it experiences a fluctuating noise with some finite temporal correlations, in this limit the previously described (standard) KPZ equation can be applied resulting in interfaces with $H = \frac{1}{2}$.

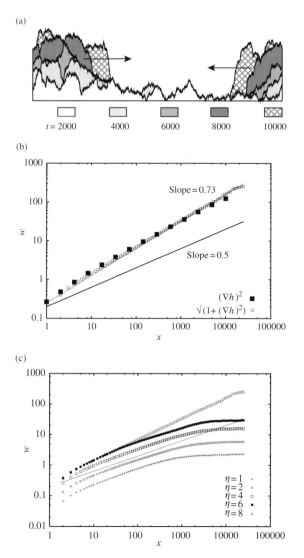

Fig. 4.17 Results of the numerical integration of the quenched KPZ equation in the noise-dominated regime. (a) Subsequent snapshots of the evolving surface. The interface consists of blocked segments and quite straight, moving parts marked by arrows. (b) The surface width w vs the length x over which its average was calculated. The straight part of the double-logarithmic plot indicates self-affine scaling with an exponent $H = 0.73$. The universality of H is demonstrated by simulations with two distinct nonlinear terms. (c) For small noise amplitudes $H = \frac{1}{2}$ is recovered (solid line). (After [21, 23, 39].)

We are, however, interested in the moving, but noise-dominated, regime. In this case the interface consists of both blocked and freely moving, almost straight, segments; thus one can expect a *higher* roughness exponent than that of the blocked interface. By extensive numerical simulations it has been shown that in this case the resulting surfaces (Fig. 4.17) have self-affine exponents ranging from 0.71 [21, 23] to 0.75

[39], which values are reasonably close to those obtained experimentally eqn (4.7). The agreement between the experimental findings and the behaviour of eqn (4.18), eqn (4.34) and eqn (4.36) is even better than this. As has been recently reported in [41,67], colonies grown on soft agarose gels (region D in Fig. 4.4(a)) – where the pinning of the interface is less dominant – indeed show standard KPZ-like behaviour characterized by $H = \frac{1}{2}$.

4.1.4 Branching morphology

If the experiments are performed on *nutrient-poor* agar substrates, then complex, branching colonies develop as first reported by Fujikawa and Matsushita [28,42]. In their study a few weeks' growth of a *Bacillus subtilis* colony resulted in beautiful branching structures as demonstrated in Fig. 4.18.

Similar patterns (Fig. 4.19) were observed in experiments with *Citrobacter*, *Escherichia*, *Klebsiella*, *Salmonella*, *Serratia* and *Proteus* strains as well [44]. According to a recent extensive study [55], many strains belonging to the *Bacilli* species also display tip splitting growth. For a – probably not complete – list of the reported strains involved in branching colony formation see Table 4.1, which demonstrates that branching colony formation, though not exhibited by *all* the strains, is really not rare under nutrient-deficient conditions.

Many of the reported findings can be explained by assuming that the growth of the colony is *diffusion-limited*, e.g., the multiplication of the bacteria is determined by the locally available amount of nutrients. At the beginning of colony development, the local nutrient supply is sufficient to maintain the growth. After further bacterial multiplication, however, nutrient deprivation will progress in and around the colony.

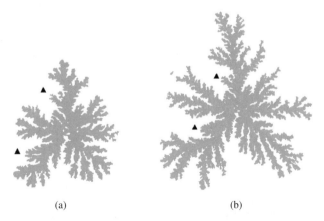

(a)   (b)

Fig. 4.18 The growth of a *Bacillus subtilis* colony, under nutrient-poor conditions. The pictures were taken 8 (a) and 19 (b) days after inoculation. Many inner branches, typical examples of which are pointed to by black triangles, are seen to stop growing (after [42]). Copyright 1990 reprinted with permission from Elsevier Science.

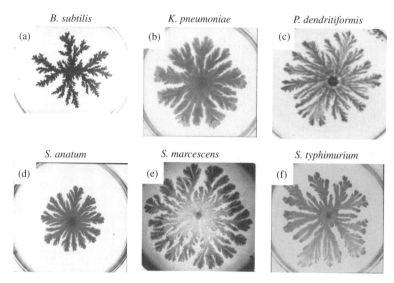

Fig. 4.19 A few example of strains exhibiting tip-splitting growth. (a,e: after [45]; b,d: after [44]; c: after [16]; f: after [43]).

Thus, bacterial growth will depend on the diffusive transport from distant regions of the Petri dish. This assumption was supported by experiments with nonmotile *B. subtilis* bacteria seeded into strongly inhomogeneous initial nutrient distribution [42]. If the nutrient was localized in a certain region of the agar gel, then growth occurred towards that region only (Fig. 4.20). Since the particular cells were nonmotile, this directed growth could not be a chemotactic effect.

Mullins-Sekerka instability in diffusion-limited systems

Diffusion-limited pattern formation has been extensively studied in nonliving systems (Fig. 4.21) because of its relevance in processes like crystal growth, electrochemical deposition or viscous fingering (e.g., see, [65]). The growth of the objects in diffusion-limited systems is determined by an external field u (e.g., the concentration) as

$$v_n = n\nabla u, \tag{4.41}$$

where v_n and n denote the dimensionless velocity and the normal vector of the surface. The dynamics of the field is governed by the diffusion equation

$$\partial_t u = D\nabla^2 u \tag{4.42}$$

with boundary conditions $u(x \rightarrow \infty) = u_\infty = \text{const}$ and $u(x) = u_\Gamma(x)$ at the interface. This latter boundary condition can incorporate some features of the microscopic dynamics, e.g., the anisotropy of the crystal structure. For the most simple, isotropic case u_Γ can be written in the form of

$$u_\Gamma = d_0\kappa, \tag{4.43}$$

Table 4.1

Bacterial strains developing branching colonies under nutrient-deficient conditions. BP: Bacto-Peptone agar, ≈ 1 g/l peptone; VB: Vogel-Bonner agar, $\approx 0.4\%$ glucose; BTB: bromthymol blue agar; TS: tryptic soy agar; NAS: soft nutrient agar, $\approx 0.4\%$ glucose. *As suggested in [55], these *B. subtilis* strains probably coexist with group I *Bacilli* which are responsible for the tip-splitting growth.

Bacterial strain	Substrate	Reference

Bacillus alvei
  PB1875
  NRRL B-383
	BP	[55]

Bacillus amyloliquefaciens H	BP	[55]

Bacillus circulans
  ATCC 4513
  NRRL B-378
	BP	[55]

Bacillus licheniformis FDO1	BP	[55, 70]
Bacillus macerans BKM B-51	BP	[55]
Bacillus popillae NRRL B-2309	BP	[55]

Bacillus subtilis
  OG-01
  W23,BD170,BD79,SB25,IS58,
  IS56,YS1,RM125,OI1085,
  OI2836,OI1055,OI3180 *
	BP
	BP
		[28, 42, 46]
		[55]

Citrobacter freundii NPC 3003	BTB	[44]

Escherichia coli
  ATCC 25922
  W 3110
	VB
	VB
	TS
		[44]

Klebsiella pneumoniae Ful	VB	[44]
Paenibacillus dendritiformis 'T morphotype'	BP	[14, 61]
Proteus mirabilis NPC 3007	NAS	[44]

Salmonella anatum
  KS 200
  KS 700
	VB
	VB
		[44]

Salmonella typhimurium
  LT2
 	
  ATCC 14028
	VB
	BTB
	VB
		[44]

Serratia marcescens
  NS 25
  NS 38
	VB
	VB
		[44]

Fig. 4.20 A colony of *Bacillus subtilis* showing a clear tendency to grow toward the nutrient (peptone), which was initially put at the right half of the dish (after [42]). Copyright 1990 reprinted with permission from Elsevier Science.

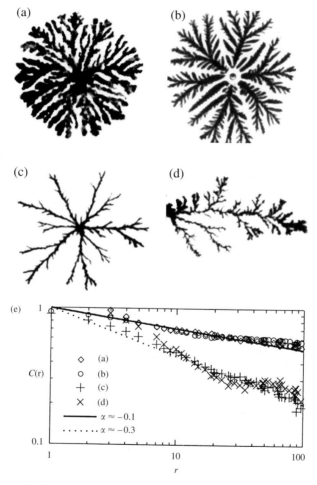

Fig. 4.21 Irrespective of the underlying physical system, the morphologies resulting from diffusion-limited growth are very similar. The dense branching morphologies ((a) crystallisation in amorphous $Al_{0.4}Ge_{0.6}$ thin films at 230°C, after [10]; (b) viscous fingering in liquid crystal, after [19]) have a fractal dimension $D \approx 2$. In the case of Laplacian growth ((c) viscous fingering in a non-newtonian liquid where the viscosity decreases with the shear [40]; and (d) crystallization in an amorphous $GeSe_2$ thin film at 220°C, after [53]) a sparse object develops with $D \approx 1.73$.

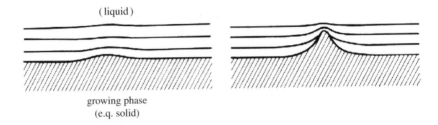

(liquid)

growing phase
(e.q. solid)

Fig. 4.22 The Mullins–Sekerka instability. If a bump is created by some fluctuation, then the gradient of the field is larger at the tip of the bump, and so the tip's growth will be further enhanced.

where d_0 is proportional to the capillary tension and κ is the curvature of the interface.[1] If the growth of the cluster is so slow that the field u can be considered stationary for a given configuration of the cluster, then we arrive at the so-called Laplacian growth. In such cases the field satisfies the Laplace equation

$$0 = \nabla^2 u. \tag{4.44}$$

Although eqns (4.41) and (4.42) are linear, the moving boundary condition for u at the interface makes the problem rather hard to study both analytically and numerically.

To show that the interface is *unstable* against perturbations [48] (Fig. 4.22), let us consider a growing circle of radius R in a diffusive field with a boundary condition eqn (4.43) and $u(R_\infty) = 1$, where $R_\infty \gg R$. The field $u(r, \phi) = u_0(r)$ satisfies eqn (4.44), which reads in polar coordinates as

$$u_0'' + \frac{u_0'}{r} = 0, \tag{4.45}$$

yielding

$$u_0 = 1 + B \ln \frac{r}{R_\infty}, \quad \text{with } B = \frac{d_0/R - 1}{\ln R/R_\infty} > 0, \tag{4.46}$$

and the velocity of the interface is

$$\frac{dR}{dt} \equiv V(R) = u_0'(R). \tag{4.47}$$

After [38], the linear stability analysis of this system will be performed in the *quasistationary* approximation (for the general solution see, for example, [37, 56]). If the growing surface is perturbed as $R(t) = R_0(t) + R_1(t, \phi)$, then u is changed as well. In general, eqn (4.42) gives rise to memory effects: a deformation of the interface causes a perturbation of the diffusion field which, in turn, affects the motion of the interface at later times. However, if the interface moves slowly enough, then it remains

---

[1] As we will show, this dependence of the local curvature is needed to avoid the formation of arbitrary thin needles in the system. Instead of eqn (4.43), one could also introduce a constraint limiting the maximal curvature of the interface.

effectively stationary during the time needed for the relaxation of the diffusion field. Thus it is reasonable to solve the problem approximately by, first, solving the Laplace equation eqn (4.44) with the appropriate boundary conditions, and then inserting this field into eqn (4.43) to obtain the time development for R_1.

Now we write the perturbation in the form of

$$R_1(\phi, t) = \hat{R}_1(t) \cos m\phi, \tag{4.48}$$

$$u(r, \phi, t) = u_0(r, t) + \hat{u}_m(t)u_m(r, \phi), \tag{4.49}$$

with $\hat{R}_1 \ll R_0$, $\hat{u}_m \ll 1$ and $u_m = r^{-m} \cos m\phi$. Both u_0 and u_m are solutions of eqn (4.44) and with this particular form the boundary condition eqn (4.43) can be also satisfied. The curvature κ in polar coordinates is given as

$$\kappa = \frac{R^2 + 2R'^2 - RR''}{(R^2 + R'^2)^{3/2}} \approx \frac{1}{R_0} - \frac{R_1''}{R_0^2}, \tag{4.50}$$

where now the prime denotes derivation with respect to ϕ. Thus, in a linear approximation eqn (4.43) reads as

$$d_0 \left(\frac{1}{R_0} + \frac{\hat{R}_1 m^2 \cos m\phi}{R_0^2} \right) = u(R(\phi), \phi), \tag{4.51}$$

where

$$u(R(\phi), \phi) \approx u(R_0, \phi) + (\partial_r u)(R_0, \phi)\hat{R}_1 \cos m\phi, \tag{4.52}$$

resulting in the following relation between $\hat{R}_1$ and $\hat{u}_m$.

$$\left(\frac{m^2 d_0}{R_0^2} - V(R_0) \right) \hat{R}_1 = \frac{\hat{u}_m}{R_0^m}. \tag{4.53}$$

Substituting eqn (4.53) back into eqn (4.41) and keeping the linear terms of $\hat{R}_1$ one finds

$$\frac{d\hat{R}_1}{dt} = \Lambda_m(R_0)\hat{R}_1 \tag{4.54}$$

with

$$\Lambda_m(R_0) = \frac{m}{R_0} V(R_0) + V'(R_0) - \frac{m^3 d_0}{R_0^3}. \tag{4.55}$$

Since Λ_m is the amplification rate of the perturbation, its sign determines the stability of the interface in respect to perturbations characterized by the m/R_0 wavenumber. For $\Lambda > 0$, the perturbation grows exponentially in time, while it dies out quickly if $\Lambda < 0$. It can be seen from Fig. 4.23 that there is a region where the amplification rate is positive. The upper cutoff is due to the surface tension represented by d_0. If $d_0 = 0$, the growth rate increases indefinitely for arbitrary short wavelengths, and the problem is ill-defined from the physical point of view.

Finally, we expect the dynamics to select the curvature for which Λ is maximal. Thus, to obtain tip-splitting growth the particular shape of the Λ_m curve is not important until it has a positive maximum value at some finite wavenumber, which can explain the frequent occurrence of the tip-splitting morphology.

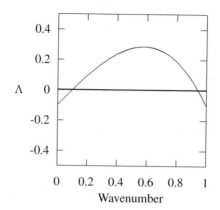

Fig. 4.23 The growth rate Λ of a perturbation vs the wavenumber m/R_0 for $V = 1$, $V' = 0.1$ and $d_0 = 1$. Deformations with $\Lambda > 0$ grow in an unstable manner. We expect that the unstable interface is characterized by the curvature for which Λ is maximal.

Diffusion-limited aggregation

The first model reproducing the branching fractal clusters characteristic for Laplacian growth is due to Witten and Sander [71]. Their model (DLA) is somewhat similar in spirit to the Eden model: particles released one-by-one perform random walks until absorbed by a growing, nonmoving cluster. When a particle is absorbed, far from the cluster a new particle is released. A typical example of the clusters obtained from computer simulations is shown in Fig. 4.24. As the power-law decay of the two-point correlation function indicates, the DLA clusters are scale invariant and are characterized by the fractal dimension

$$D_{\text{DLA}} = 1.715. \tag{4.56}$$

It is easy to see that the DLA model is indeed related to the diffusion-limited processes described above. Let us denote by $p(x, y)$ the probability of the event that – during a long enough time – a given particle passes the lattice site (x, y). If the cluster did not absorb the particles, then $p(x, y)$ would be constant everywhere. Since the particles are absorbed at the surface, inside the cluster $p = 0$, while at large distances from the cluster $p = constant$ holds. The growth probability of the various parts of the cluster is determined by p near the boundary. Since at the boundary $p = 0$, on average the growth is proportional to ∇p at the surface. Finally, the stationary distribution of p satisfies the Laplace equation, as shown by the local balance equation of the lattice version of the DLA model relating $p(x, y)$ to p at the neighbour sites as

$$p(x, y) = \tfrac{1}{4}[p(x - 1, y) + p(x + 1, y) + p(x, y - 1) + p(x, y + 1)], \tag{4.57}$$

since the probability of stepping in any direction is $\frac{1}{4}$.

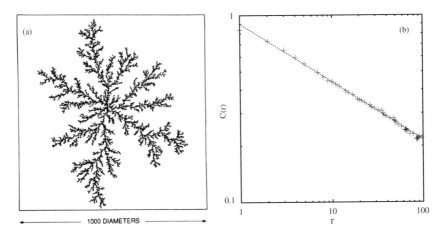

Fig. 4.24 (a) Typical DLA cluster with 50,000 particles, reproduced from [47]. (b) The object is self-similar with $D \approx 1.7$, as the power-law decay (dotted line) in the density–density correlation function shows.

Many theoretical attempts has been made to understand how the nontrivial fractal structure of Fig. 4.24 emerges from the simple models of diffusion-limited growth. Although there is no established theory yet, the most promising approaches are based on some sort of *renormalization transformation*, e.g., see [69].

Many natural processes can be represented better by growing clusters from a pool of random walkers that initially occupy a finite fraction p of the lattice. In the numerical simulations the particles are selected randomly and moved to a nearest-neighbour site. If the new site is on the perimeter of the growing cluster, then the particle becomes a part of the cluster.

The clusters generated by this model also have DLA-like structure on short length scales $1 \gg r \gg \xi$ and are uniform for $\xi \gg r$ (Fig. 4.25). The crossover length ξ is determined by imposing the condition that the average density of the cluster and that of the pool of particles must be the same. At the same time ξ is also the width of the depleted region around the growing pattern. Thus, the density–density correlation function eqn (4.1) has the form

$$C_\varrho = \begin{cases} r^{-\alpha} & \text{for } r \gg \xi \\ \text{const} & \text{for } \xi \gg r. \end{cases} \tag{4.58}$$

Models resolving individual bacteria

To understand the diversity of bacterial patterns and their microbiological regulation, more elaborated models are needed than the previously presented continuum description of the diffusion-limited growth or the DLA model. We have two major classes of methods at our hand, both with advantages and shortcomings: the continuum description by differential equations and the models based on the simulation of individual organisms/particles.

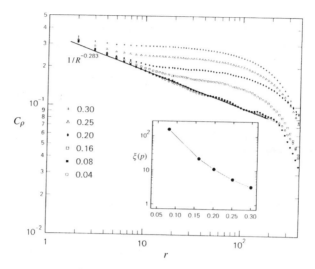

Fig. 4.25 Density–density correlation function for multi-particle DLA aggregates at different concentrations p of the diffusing particles. At low values of p the standard DLA fractal behaviour is recovered (solid line). The inset shows how the correlation length ξ varies with p (after [66]).

As we have seen, when working with differential equations a coarse-grained description is given which is valid for larger length scales than that of the microscopic cell–cell interactions. Hence, in this framework it is usually not straightforward to find the most suitable formulas representing the various basic microbiological processes. However, in some cases these models might be treated analytically yielding valuable information on the robustness and universality of the description.

The other way is to model the *individual* bacteria by some idealized units, defining their behaviour and interactions, and obtain the large-scale behaviour of the system by computer simulations. Models for single bacterial cells have been published for *E. coli* [24, 32] and *B. subtilis* [31]. The purpose of these studies was to integrate subcellular processes into whole-cell models, involving the use of much information on the biochemistry of these bacteria. However, in our case a much simpler description is sufficient, since bacterial growth on the cellular scale can be described by rather simple dynamics [35]. When modelling individual units, it is convenient to incorporate into the model details like the cell cycle or the differentiation of the cells, allowing the exploration of more sophisticated behaviour. At the same time such models are not really suitable for analytical treatment.

Now we focus on some variants of the individual-based models, originally introduced in [16, 17] and – as we show in the next section – later successfully applied for other colony formation processes as well. Such models are constructed as *minimal* models with the following guidelines: both the *microscopic* 'rules' and the calculated/simulated behaviour of the whole system on the *macroscopic* level should be in accord with experimental observations. In fact, the models are often built in an iterative manner: the morphology diagram is approximated with the inclusion of more and more microscopic details, until satisfactory agreement has been reached at the macroscopic level. Of course, at the microscopic level the model always gives a poor representation of the biological entities, but based on our experience with different models in the previous sections we have a good chance that several such details are irrelevant in the pattern formation process. The resulting picture can then be verified by additional experimental or analytical studies.

Modelling nonmotile bacteria In the simplest version of such models, each particle i is characterized by its *position* x_i and *cell cycle state* E_i.

1. Changes in E_i control the *sporulation* and *division* of the cells as follows. If E_i decays below a threshold value (which can be specified as zero) then the particle becomes inactive (sporulates)[1]. Above another threshold value, say 1, the model bacterium divides, and both of the daughter cells receive an initial value of $0 < E_* < 1$. It is known [52] that long-lasting starvation initiates the sporulation process, and the average multiplication rate is influenced by the available amount of nutrients. Thus, in the model, changes in E_i are coupled to the nutrient consumption rate ω_i as

$$\frac{dE_i}{dt} = \kappa \omega_i - \epsilon, \tag{4.59}$$

where κ is a conversion factor relating the maximal nutrient consumption rate to the shortest cell cycle time, and ϵ is the generic 'maintenance' term incorporating all free-energy expenditures not directly contributing to growth [36, 62].

2. The *nutrient uptake* is limited by both the local concentration c and enzymatic rates. It is approximated by a Michaelis–Menten-like kinetics (see Fig. 4.3) as

$$\varrho(x_i)\omega_i = \min \left[\omega_{max} \varrho(x_i), \omega_0 c(x_i) \right], \tag{4.60}$$

where ϱ is the local cell density, ω_{max} is the maximal uptake rate of the cells and $\omega_0 c$ is the maximal diffusive transport from the substrate to the cells.

3. Changes in c are given by the diffusion equation with the appropriate sink terms at the position of the active particles as

$$\frac{\partial c}{\partial t} = D_c \nabla^2 c - \sum_i \omega_i \delta(x - x_i). \tag{4.61}$$

By appropriate choice of units, $\kappa = 1$ and $\omega_{max} = 1$ can be assumed. Thus, the remaining control parameters of the model are E_*, ω_0, ϵ, D_c, and the initial nutrient concentration c_0. As the simplest choice, here we investigate the $E_* = 1/2$ and $\epsilon = 0$ case, i.e., when all the consumed nutrient is transformed into biomass. We fix $\omega_0 = 0.1$

---

[1] In all of the examples considered here this process is practically irreversible, since sporulation occurs deep inside the colony, where the nutrient-deficient condition remains through the colony growth.

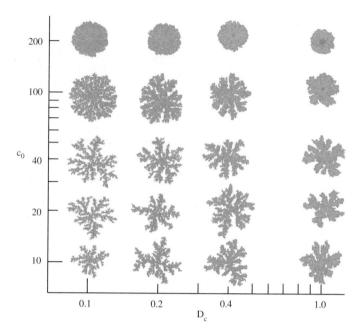

Fig. 4.26 Morphology diagram generated by the model with nonmotile particles as a function of the initial nutrient concentration c_0 and nutrient diffusivity D_c. The colonies were grown (in the computer) until either their size or the number of bacteria reached a threshold value.

and systematically explore the parameter space with respect to D_c and c_0. Note that in the experiments c_0 is directly controllable while D_c is determined by the hardness of the agar substrate. Since the colonies can consist of up to 10^{10} cells, even with the present computational power it is impossible to treat that many units. Thus, we might think of the simulated entities as if they represent about 0.1% of the total population.

Figure 4.26 shows the morphology diagram obtained by numerical simulations. When the food concentration is high, the resulting patterns are compact, while they become branching for more hostile conditions in agreement with the experimentally determined morphology diagram of the non-motile *B. subtilis* OG-01b strain (Fig. 4.4(b)). The calculation of the C_ϱ density–density correlation function (4.1) reveals (Fig. 4.27) that the scaling characteristic for diffusion-limited objects ($D \approx 1.7$) is limited by both an upper cutoff determined by the finite diffusion length of the nutrients, similar to the case of multi-particle DLA (Fig. 4.25), and a lower cutoff due to the finite width of the branches. This latter quantity increases with D_c as one can see from the following simple argument. Inside the colony the cell density is high and the nutrient uptake is approximately $\omega_0 c$. Therefore, c decreases as

$$c \sim e^{-z\sqrt{\omega_0/D_c}}, \tag{4.62}$$

with z being the distance measured from the colony boundary. Since the growing branches have a stationary shape, the infiltration depth of the nutrients, $\sqrt{D_c/\omega_0}$, and the width of the branches are proportional.

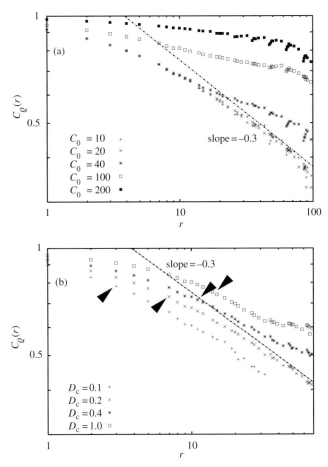

Fig. 4.27 Density–density correlation functions calculated for the morphologies shown in Fig. 4.26. (a) When increasing the initial nutrient concentration c_0 at $D_c = 0.2$, the DLA-like fractal behaviour ($\alpha = 0.3$, lines) changes to a compact morphology (compare with Fig. 4.25). (b) For a given c_0 ($c_0 = 20$) one can observe the parallel increase of the nutrient diffusivity D_c and the lower cutoff length (marked by arrowheads) corresponding to the width of the branches.

This behaviour of $C_\varrho(r)$ also explains the lack of tip-splitting growth in certain colonies. If the penetration depth of the nutrients is too large or the diffusion length outside the colony is too small, then the upper and lower cutoff values can be arbitrarily close, which results in compact growth within the experimentally accessible range of parameters.

Modelling motile bacteria   Now let us turn our attention to the morphology diagram exhibited by *P. dendritiformis* (Fig. 4.5). As microscopic observations revealed, in this case the bacteria are motile. However, they still cannot migrate on dry agar gel surfaces, but only where some surfactant was secreted (Fig 4.28). To explore this

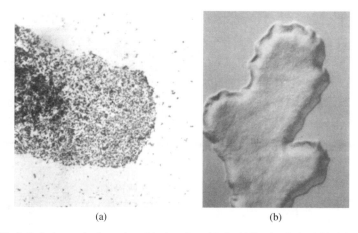

(a)   (b)

Fig. 4.28 Optical microscopic observation of the branches of *P. dendritiformis* colonies. (a) Staining reveals the short, rod-like bacteria in the growing tips of the colonies. The stronger-stained objects behind the tip are spores. (b) Interference microscopy using Nomarsky optics reveals difference in optical depth due to secreted extracellular fluid. (After [17].)

system, we keep the rules (a)–(c) of the previous model, but also include new ones describing the motion of the bacteria.

(d) The active particles move randomly within a *boundary* described by

$$\frac{\mathrm{d}x_i}{\mathrm{d}t} = v_0 e, \tag{4.63}$$

where e denotes a random vector of unit length and uniform angular distribution.

(e) The propagation of the boundary is assumed to be proportional to the local density of active bacteria. In fact, the collisions of the particles with the boundary were counted and when a threshold value (N_c) was reached the new neighbouring cell could be occupied. This rule mimics the surfactant or extracellular slime secretion and deposition in a crude fashion.

The resulting morphology diagram is shown in Fig. 4.29. Figure 4.29 certainly resembles the morphology diagram in Fig. 4.5 within a limited region of the parameters, but it fails to predict the formation of the thin, straight radial branches and the corresponding increase in the fractal dimension at very low food concentrations (Fig. 4.6). To explain these structures a repulsive chemotaxis signalling mechanism was suggested in [17]. It was assumed that the bacteria which are in the sporulation phase emit certain diffusive chemicals acting as *chemorepellents*, i.e., the random swimming of the bacteria becomes biased: the cells move less toward the chemical gradient (Fig. 4.30).

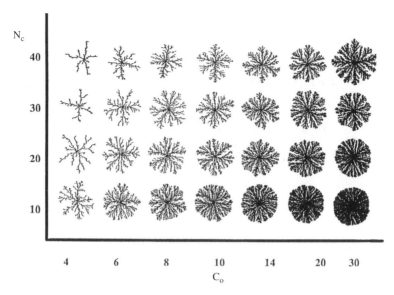

Fig. 4.29 Morphology diagram generated by the model with motile bacteria as a function of the initial nutrient concentration and agar gel 'hardness', i.e., the threshold value for the colony borderline displacement (after [17]). Reprinted by permission from Nature copyright 1994 Macmillan Magazines Ltd.

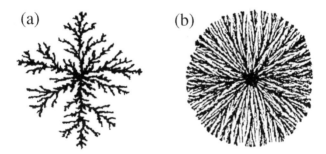

Fig. 4.30 Effect of repulsive chemotaxis signalling in the model with motile bacteria. Results are shown of simulations without (a) and with (b) chemotaxis signalling. The pattern becomes denser and shows radial thin branches and a well-defined circular envelope in agreement with experimental observations (after [17]). Reprinted by permission from Nature copyright 1994 Macmillan Magazines Ltd.

Additional studies [20] with such models revealed a further possible role of bacterial chemotaxis signalling in the colony development. As Fig. 4.31 demonstrates, *chemotaxis towards nutrients* does not change the morphology but significantly enhances the colony growth. Thus, based on such investigations and experimental studies of the mixed morphologies shown in Fig. 4.6, it was suggested that different chemotactic effects dominate the growth at various parts of the morphology diagram: while above a certain threshold nutrient concentration ($c_* \approx 1$ g/l) the food chemotaxis is relevant, below that concentration the repulsive signalling is the main factor affecting the morphology. Furthermore, since these changes in the morphology are not inheritable, it

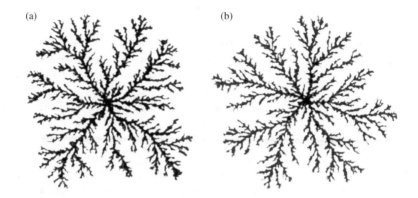

Fig. 4.31 The effect of food chemotaxis in the same model. Results of simulations without (a) and with (b) chemotaxis signalling. In the latter case the growth velocity is twice as large as that of (a), yet the patterns have the same morphology (after [20]). Copyright 1996 reprinted with permission from Elsevier Science.

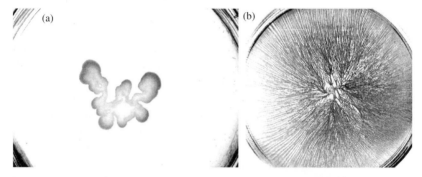

Fig. 4.32 Chemotaxis deficient mutant of *S. marcescens* (a) versus the wild type (b). Note that the dense branching morphology disappears (after [43]).

was assumed that below c_* the food chemotaxis 'shuts down'. Unfortunately, molecular biological studies on the chemotaxis system of the *P. dendritiformis* strain have not been performed yet. However, experiments with another strain, *S. marcescens*, support the hypothesis that chemotaxis is really involved in the generation of dense radial branching patterns (Fig. 4.32).

4.1.5 Chiral and rotating colonies

When bacterial motion is strongly influenced by cell–cell interactions, i.e., if the cells cannot be treated as random walkers with a bias towards some external concentration gradients, then the colony formation process can be even more complex than that shown in the previous sections. Here we give some examples for such a behaviour and demonstrate how the modelling approach can help the understanding of these systems.

Chiral patterns

Under certain conditions the strain *Paenibacillus dendritiformis* can *reproducibly* change its behaviour, resulting in a morphology diagram different from that shown in Fig. 4.5. As described in [14,61] and independently verified in [55], after a growth of about two days on soft substrates (agar concentrations below 10 g/l), a sudden burst of branches with a new, *chiral* morphology can be observed: each branch is curved and has the same handedness (see Fig. 4.33). New colonies inoculated with bacteria from these branches exhibit chiral morphology without any adaptation period. The transition of growth morphology is reversible: after some incubation on harder substrate the chiral growth reverts back to the much-studied tip-splitting morphology. Therefore, at agar concentrations around 10 g/l, the developed patterns are not unambiguously determined by the current environment, but also depend upon the conditions experienced by the bacteria in the past. Recent molecular biological [55] and microbiological classification studies [61] failed to find any difference between the bacteria taken from the two types of colonies; hence they are assumed to be two distinct phenotypes of the same strain.

Albeit less frequent than tip-splitting patterns, the chiral morphology of *P. dendritiformis* is not unique: it has been reported for certain *Bacillus subtilis, Bacillus macerans* [55] and *Serratia marcescens* [43] strains as well. As the same handedness develops in an upside-down grown colony [16], the chirality is not an effect of some external (e.g., magnetic or Coriolis) force. When looking for possible mechanisms for the chiral symmetry-breaking in the case of *P. dendritiformis*, microscopic observations revealed that the cells are very elongated and their orientation is parallel to each other (Fig. 4.34). Furthermore, as described in [13], the motion of the bacteria is strongly correlated with their orientation: they move approximately parallel to their axis. Already this highly coordinated *gliding* motion of the cells shows a certain level

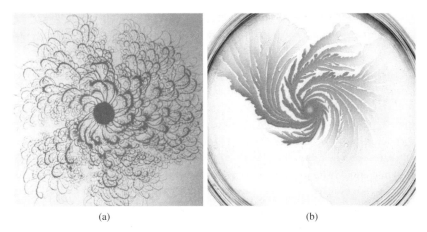

(a)   (b)

Fig. 4.33 Chiral growth of bacterial colonies. (a) The chiral form of *Paenibacillus dendritiformis* on soft agar (after [15]). (b) Serrawettin-deficient mutant of *Serratia marcescens* (after [43]).

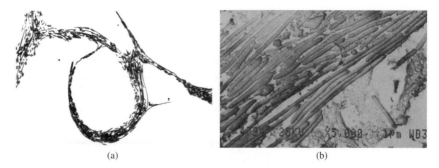

(a)   (b)

Fig. 4.34 Optical (a) and electron- (b) microscope images of the chiral form of *P. dendritiformis*. The bacteria are long and exhibit a high level of orientational correlations (after [16]).

of chirality. In addition, individual bacteria can leave a 'trail' which can guide the motion of the following cells. Less microscopic detail has been published for the other strains exhibiting chiral morphology. The *S. marcescens* strain is known to be deficient in producing serrawettin, a surfactant agent. In that case the lack of sufficient lubrication layer may enhance the coordination between the individual cells and give rise to a more similar microscopic behaviour than that of the chiral form of *P. dendritiformis*.

The effect of the above features on the colony morphology was explored with model calculations [13]. In the simulations the particles were characterized not only by their location x_i or cell-cycle state E_i but by their orientation ϕ_i as well. The food consumption, cell activity and colony boundary advancement rules were the same as described in the previous section. However, the motion of the particles was not a biased or unbiased random walk; instead, their displacement was parallel to their orientation ϕ_i. Finally, ϕ_i was changed by two factors: the intrinsic chirality of the motion and the alignment to a local orientation field Φ representing neighbouring cells and trails on the surface. In fact, the time development of ϕ_i was defined as

$$\phi_i(t + \Delta t) = P(\phi_i(t), \Phi(x_i(t))) + \varphi + \xi_i(t), \qquad (4.64)$$

where φ represents the intrinsic chirality of the motion of the bacteria, ξ is a random noise with an amplitude less than $\hat{\xi}$ and the interaction with the local orientation field is described by the projector function P as $P(\alpha, \beta) = \alpha + (\beta - \alpha) \bmod \pi$. The value of Φ was set at the occupation of the particular lattice site and remained unchanged. Since Φ represents the tracks in the agarose gel, its initial value was chosen to be the orientation of the first walker entering that field.

In Fig. 4.35 a detailed drawing of the model is presented and results of the numerical simulations with various parameter values are shown, demonstrating that the microscopic chiral behaviour can be indeed amplified to macroscopic scales. In some sense it is a similar phenomenon to the appearance of the hexagonal crystal structure of ice in the snowflake patterns and it is strongly dependent on the preservation of the orientational correlations among the adjacent particles. If the noise amplitude $\hat{\xi}$ is increased, then the chiral nature of the pattern disappears and tip-splitting growth

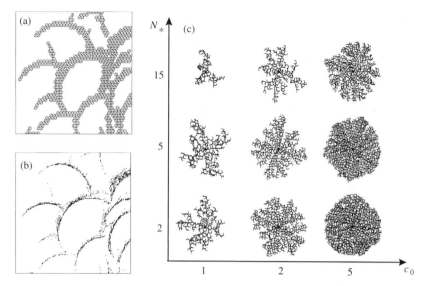

Fig. 4.35 Computer simulation of chiral colony formation. (a) The occupied lattice sites with their orientation. (b) The position and orientation of the particles within the colony. (c) Colony shapes for various values of the initial nutrient concentration c_0 and agar hardness N_c (after [13]). Copyright 1995 by the American Physical Society.

takes place. These results are in complete agreement with experimental observations. Bacteria are shorter and less aligned when grown on nutrient-rich substrates, where the chirality of the colonies is indeed lost [13].

Self-generated vortices

The rotation of disc-shaped aggregates consisting of a few or millions of bacteria, or their peculiar motion along ring-shaped trajectories, has been reported long ago [27, 60,72]. Although such types of motion have commonly been associated with *Bacillus circulans*, similar phenomena have been seen in colonies of *Archangium violaceum, Chondromyces apiculatus, Clostridium tetani* and in colonies of the recently identified strain of *Paenibacillus vortex* [12, 18] as well.

In Fig. 4.36 a typical *P. vortex* colony is shown. As described in [11,22], the bacteria spread via droplets (the darker dots in the figure), leaving a trail mostly filled with spores behind. Each droplet consists of many cells moving at the perimeter with typical velocities of 10 µm/s (Fig. 4.37). Depending on the growth conditions and the location inside the colony, the number of bacteria in a single vortex can vary from a couple of hundreds to many thousands, and the vortex can consist of both single and multiple layers of bacteria. Usually, the 'pioneering' droplets are larger, while the smaller ones fill the empty areas left behind by the advancing front. Within a single colony, both clockwise and anti-clockwise rotating vortices can be observed.

In §6.1 we show that the geometrical constraints of the boundaries can enforce a circular collective motion in a system of locally interacting self-propelled particles.

Fig. 4.36 A typical colony of the strain *Paenibacillus vortex*. Each branch is formed by a rotating droplet of many bacteria moving together in a correlated manner at the branch tips (after [22]). Copyright 1996 by the American Physical Society.

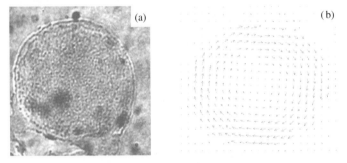

Fig. 4.37 Bright field micrograph of a single rotating droplet with a magnification of 500× (a) and the corresponding velocity field obtained by digitizing the videomicroscopic recordings (b) (after [22]). Copyright 1996 by the American Physical Society.

This result suggests that a radial inward force acting on the migrating bacteria can explain the observed vortices.

Since collectively migrating bacteria do not 'tumble', their response to chemo-tactic agents must differ from that of swimming cells. As Fig. 4.38 demonstrates, such a chemotactic effect certainly exists, although its molecular bases are not yet known. Thus, in the following part of this section we use the words 'chemotaxis' and 'chemoattractant' in a broad sense, the particular response we consider can easily result from passive physical forces like surface tension or changing efficiency of translocation with the amount and quality of the extracellular slime deposited. In [22] such a special kind of chemotactic response was proposed: if the propulsion force of the individual cells is modulated depending on the local chemical concentrations, then it results a torque acting on the whole group.

To see this, let us consider a group of bacteria moving as a rigid body in a concentration field c_A. In a linear approximation the propulsion forces F are distributed as

$$F(r) = F_0(1 + ar\nabla c_A) \qquad (4.65)$$

where F_0 is the average propulsion force, r is measured from the center of the group and a is a coefficient. Since at the microscopic length scales the inertial effects are

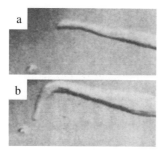

Fig. 4.38 'Purposeful movement' of *Myxococcus xanthus* cells is shown in consecutive frames with 7 minutes time difference. The cells collectively migrate toward a latex bead at the bottom of the images (after [57]).

negligible, for the velocity v of the whole group one can assume

$$v \sim \int F(r)\, \mathrm{d}^2 r \sim F_0 , \qquad (4.66)$$

while for its rotation

$$\omega \sim \int [r \times F(r)]\, \mathrm{d}^2 r \sim F_0 \times \nabla c_A \qquad (4.67)$$

holds. Since the velocity and the propulsion forces are strongly correlated with the orientation of the cells (see the previous section),

$$\frac{\mathrm{d} F_0}{\mathrm{d} t} \sim \frac{\mathrm{d} v}{\mathrm{d} t} = \omega \times v , \qquad (4.68)$$

and for the equation of motion of a single coherent bacterial group we obtain

$$\frac{\mathrm{d} v}{\mathrm{d} t} \sim -v \times (F_0 \times \nabla c_A), \qquad (4.69)$$

which expression is independent of the size of the group [22].

To study the interactions of the bacteria and understand the spectacular self-organized vortices, we expand eqn (4.69) to include the orientational interaction between the neighbouring cells and also to add a term proportional to the cell density gradient preventing the accumulation of walkers into arbitrary dense spots as

$$\frac{\mathrm{d} v_i}{\mathrm{d} t} = \mu \left(\langle v \rangle_{i,\epsilon} - v_i \right) - \nu v_i + F \frac{v_i}{v_i} - \kappa \nabla \varrho$$
$$- \frac{\chi_A F}{v_i} v_i \times (v_i \times \nabla c_A) + \zeta. \qquad (4.70)$$

The first term on the right-hand side of eqn (4.70) incorporates both the orientational interaction and the momentum transfer between the neighbouring cells (see §6.1). The $\langle v \rangle_{i,\epsilon}$ average is calculated over the particles within a distance ϵ from the walker i. The next two terms express the self-propulsion and the overdamped nature of the motion. The last term in eqn (4.70) is an uncorrelated noise.

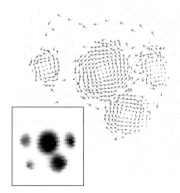

Fig. 4.39 Typical result of model simulations for vortex formation. The positive chemotactic feedback breaks the originally homogeneous density resulting in aggregation of the cells. The flow field is represented by arrows of a magnitude proportional to the local velocity. The inset shows the concentration distribution of the chemoattractant. (After [22].) Copyright 1996 by the American Physical Society.

The time evolution of the diffusing chemoattractant is described by

$$\frac{\partial c_A}{\partial t} = D_A \nabla^2 c_A + \Gamma_A \varrho - \lambda_A c_A. \tag{4.71}$$

The first and the third terms on the right-hand side represent the diffusion and a constant rate decay, respectively, while the second (source) term assumes that bacteria produce the chemoattractant at a constant rate. This is a useful assumption if we investigate the formation of a single vortex, but must be refined when we also intend to describe the morphogenesis of the whole colony.

The simulation of the above model was carried out through the numerical integration of eqns (4.70) and (4.71), as described in [22]. The *positive feedback* of the attractive chemotaxis breaks the originally homogeneous spatial distribution of the particles and creates dense aggregates. If a fluctuation increases the density locally, then the emission of chemoattractants (being proportional to ϱ) is also increased. The increased concentration gradient attracts even more particles to the aggregate. As a consequence of the local alignment interaction, rotation develops in a spontaneously selected direction. In such an aggregate the difference of the $\nabla \varrho$ term and the attractive force of chemotaxis provides the appropriate centripetal acceleration. Typical vortices formed in the system are shown in Fig. 4.39, being in a good agreement with experimental observations (Fig. 4.37). With slightly different parameter values the model yields rotating rings reported also in experiments (Fig. 4.40). The qualitative results discussed here turned out to be insensitive to the exact form of eqn (4.70) until a cohesive force with co-alignment interaction is represented.

Thus, the model defined through eqns (4.70) and (4.71) explains many experimentally observed features in the colonies of *P. vortex*, *B. circulans* and related strains such as collective migration of groups or formation of rotating dense aggregates in which the direction of the spinning was selected by spontaneous symmetry breaking.

(a)　　　　　　　　　　　　　　(b)

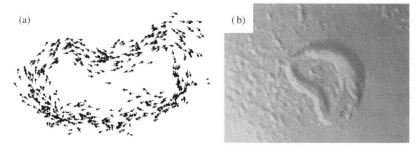

Fig. 4.40 (a) For slightly different value of the parameter μ (providing stronger velocity–velocity interaction) than that used to generate Fig. 4.39, rotating rings develop (after [22]). Copyright 1996 by the American Physical Society. (b) This phenomenon was also documented in a *B. circulans* colony [72].

Fig. 4.41 Typical colony obtained from numerical simulations of the model incorporating nutrient diffusion, reproduction and sporulation of bacteria, chemotactic regulation, velocity–velocity interaction and extracellular fluid deposition. The simulation is performed in a 600×600 system with ca. 10000 particles. In excellent agreement with the experimental observations, the colony grows via rotating droplets at the tips of the branches. Smaller vortices also emerge inside the colony, sometimes giving rise to a new side branch. In the figure the gray scale represents the density of the active bacteria (after [22]). Copyright 1996 by the American Physical Society.

However, the model is not complete since it does not describe colony formation at the macroscopic level, which would certainly be needed to understand the possible (evolutionary) benefits of this strange behaviour. To expand the model to describe colony morphogenesis, further details were considered in [22]: extracellular slime influencing the motion of the organisms, nutrient diffusion and consumption determining the growth rate of the colony, and additional chemical regulation of the movement of the individual vortices (Fig. 4.41).

The model calculations [22] also revealed an interesting benefit of vortex formation. As beyond the vortex (inside the colony) the nutrient concentration is rather small, vortices are exposed to a nutrient field with a strong gradient. Averaged over time, cells in the rotating vortices experience a less inhomogeneous concentration field; thus *both* their reproduction and sporulation rate is reduced. Therefore, by reducing their expenditure on reproductive and sporulation processes, they can spend on the average more energy to enhance the speed of their propagation. Under diffusion-limited conditions the faster expansion increases the inflow of nutrients.

The strategy of reducing the rate of multiplication and thus using the energy for colonizing surfaces is not a unique feature of vortex-forming bacteria. In the case of swarmer strains similar behaviour has been reported [6]: the differentiated swarmer cells are specialized to expand the colony, and many of their metabolic pathways (including the multiplication processes) are repressed.

The question of the *robustness* of such complex models arises in a natural way, and – according to our opinion – cannot be answered completely yet. However, one can certainly find supporting arguments. First, the experimental data for various external conditions and organization levels (ranging from possible microscopic interactions through the flow fields at intermediate length scales to features of the macroscopic colony formation) yields, in fact, strict constraints for the models. Second, some of the generic features such as nutrient consumption, multiplication or sporulation of the bacteria and their mathematical representation can be found and verified in various other systems as well. Third, according to the experience with numerical models, usually a broad range of parameters yield qualitatively the same behaviour, i.e., in many cases no 'fine tuning' of the parameters was required.

4.2 Statistical analysis of DNA sequences

T. Vicsek

In this section recent methods of analysing the information contained in DNA molecules will be discussed. This information is necessary for reproduction of organisms and is stored in the form of a sequence which can be considered as a long series of four 'letters' A, C, G and T (corresponding to the four bases in DNA molecules).

DNA molecules are the most sophisticated information databases created by nature through the dynamic process of evolution. Equally remarkable is the precise transformation of different layers of information (replication, decoding, etc.) that occurs in a short time interval. While a mechanism of encoding some of this information is understood (for example, the genetic code directing amino acid assembly, sequences directing intron/exon splicing, etc), relatively little is known about other layers of information encrypted in a DNA molecule. In the genomes of high eukaryotic organisms, only a small portion of the total genome length is used for protein coding. The role of introns and intergenic sequences constituting a large portion of the genome remains unknown. The complexity of the information content coded in DNA sequences is partly reflected in their heterogeneous structure.

In recent years sequence data have been collected through the joint effort of many research laboratories working within the framework of the Human Genome Project.

In the near future, the amount of information is expected to increase by several orders of magnitude. One major issue facing researchers is how to reliably identify protein coding segments and other functional sites in this growing DNA database. To accurately locate such functional regions in long DNA sequences serves not only a practical purpose for guiding biologists to 'useful' information in the nucleotide database. The understanding of the exact algorithm of splicing introns (or the organization principle of coding and noncoding sequences) will definitely shed some light on the biochemical and physical mechanisms involved.

It is generally accepted that the structure of DNA is very complex and heterogeneous, and cannot be described by any single mathematical model. A new science, *Computational molecular biology*, starts to crystallize out of these methods and models. The goals of this cross-disciplinary science that incorporates ideas and methods of physics, chemistry, applied mathematics, etc. were formulated recently as (i) to understand biological phenomena that involve sequences of nucleic acids and (ii) to assist molecular biology software development. With the background of statistical and polymer physics, it has been demonstrated that the idea of long-range correlations, developed originally in statistical physics to study extremely heterogeneous systems near their critical points, is well-suited to quantify the heterogeneity and mosaicism of DNA molecules.

4.2.1 DNA walk

One of the promising approaches to the interpretation of the organization of information in DNA has been the investigation of the so-called 'DNA walk'. In this recent method, intensively investigated by H.E. Staley and his group, DNA is mapped onto a process which can be regarded as a walk. Each of the four 'letters' of a DNA sequence is identified with a step in a given direction. Then, the specific features of this walk can be analysed using methods borrowed from statistical physics [1].

The idea is to look for correlations. Two series of data (X and Y) are correlated if there is a relationship between the corresponding elements of the series. In other words, if the fact that X_i (the ith element of X) is larger than the average value of X is typically accompanied by the fact that Y_i is also larger than the average of Y, then the two sets of data are positively correlated, and the corresponding correlation coefficient is a number larger than zero. No correlations result in a coefficient equal to zero. Correlations within a single sequence of data can be defined similarly. In this case one can ask how the value X_i is related to the value X_{i+j}. By comparing the two values with the average of X one can get information about the question whether two values in the data set separated by j elements are correlated.

An ordinary random walk has no long-range correlations; thus, an analysis concentrating on the nature of correlations may detect relevant differences from random sequences. It has been shown that the patches in coding regions usually have a characteristic length scale that is roughly the average protein size. Within each patch, the sequences have only short-range correlations. On the other hand, noncoding sequences have patches of multiple length scales, creating power law or long-range correlations.

One of the most relevant questions one can raise in the context of DNA sequences is the location of coding and noncoding parts in the genome. In the case where these two kinds of subsequences have different kinds of correlations we may be able to differentiate between the coding and non-coding part without any prior knowledge about the sequences.

In order to reliably determine the nature of correlations one has to use improved algorithms compared to the simplest ones. Because of the specific features of DNA sequences (they may be very patchy and exhibit trends over many bases, e.g., due to repeats, without any particular 'meaning') the DNA walk can be best evaluated using a method that gets rid of the informationless trends. The so-called Detrended Fluctuation Analysis (DFA) is a powerful tool to locate coding parts in DNA. It consists of the following steps:

1. A nucleotide sequence $\{n_i\}$ ($i = 1, 2, \ldots, L$) of length L is comprised of the base pairs A (adenine), C (cytosine), T (thymine) and G (guanine). In order to apply numerical methods to a nucleotide sequence, we first prepare seven numerical sequences $\{u_i\}$, corresponding to seven ways of mapping the original nucleotide sequence onto a 1-D numerical sequence, which we denote using standard nucleotide base codes (IUPAC).

    (a) Purine-Pyrimidine (RY) rule: if $n_i = A$ or G, then $u_i = 1$; if $n_i = C$ or T then $u_i = -1$.

    (b) $A\bar{A}$ (AB) rule: if $n_i = A$ then $u_i = 1$; in all other cases $u_i = -1$.

    (c) $T\bar{T}$ (TV) rule: if $n_i = T$ then $u_i = 1$; in all other cases $u_i = -1$.

    (d) $G\bar{G}$ (GH) rule: if $n_i = G$ then $u_i = 1$; in all other cases $u_i = -1$.

    (e) $C\bar{C}$ (CD) rule: if $n_i = C$ then $u_i = 1$; in all other cases $u_i = -1$.

    (f) Hydrogen bond energy (SW) rule: $u_i = 1$ for 'strongly-bonded' pairs (G or C); $u_i = -1$ for 'weakly-bonded' pairs (A or T).

    (g) Hybrid (KM) rule: $u_i = 1$ for A or C; $u_i = -1$ for G or T.
    The RY rule has been the most widely used rule.

2. For each numerical sequence $\{u_i\}$ compute a running sum

    $$y(n) \equiv \sum_{k=1}^{n} u_k, \quad [y(0) \equiv 0] \tag{4.72}$$

    which can be presented graphically as a 1-D landscape or 'DNA walk'.

3. Divide the entire sequence of length L into L/ℓ nonoverlapping boxes, each containing ℓ nucleotides, and define the 'local trend' in each box (proportional to the compositional bias in the box) to be the ordinate of a linear least-squares fit for the DNA walk displacement in that box.

4. Define the 'detrended walk', denoted by $y_\ell(n)$, as the difference between the original walk $y(n)$ and the local trend. Calculate the variance about the local trend for each box, and calculate the average of these variances over all the boxes of size ℓ.

A variant of the original DFA method is when we use a sliding box, in order to obtain better statistics. Specifically, we define a sliding observation box of size ℓ that starts at base pair i and ends at base pair $i + \ell$. Then we compute the least-squares linear fit $y_{i,\ell}(n) = na + b$ such that the sum of $\ell + 1$ squares for this box

$$E_{i,\min}(\ell) \equiv \sum_{n=i}^{i+\ell} \left[y(n) - y_{i,\ell}(n) \right]^2 \tag{4.73}$$

is a minimum.

Finally, we average $E_{i,\min}(\ell)$ over all positions of the observation box from $i = 0$ to $i = L - \ell$ and define the 'detrended fluctuation function' as

$$F_{\mathcal{D}}^2 \equiv \frac{1}{(L - \ell + 1)(\ell - 1)} \sum_{i=0}^{L-\ell} E_{i,\min}(\ell). \tag{4.74}$$

Using a fast algorithm for computing $F_{\mathcal{D}}^2$ and a RISC-based workstation it takes a few hours to analyse all the primate sequences in the GenBank. For sequences with power-law long-range correlations for $\ell > 10$ the detrended fluctuation can be well approximated by a power law

$$F_{\mathcal{D}}(\ell) \sim (\ell + 3)^\alpha. \tag{4.75}$$

Thus, the exponent α can be accurately measured as the slope of a double-logarithmic plot of $F_{\mathcal{D}}(\ell)$ versus $\ell + 3$ even for very small values of ℓ.

Using DFA all 33301 coding and all 29453 noncoding eukaryotic sequences, each of length larger than 512 base pairs in the GenBank release of 15 August 1994, have been analysed. The results for a selected sequence are presented in Fig. 4.42

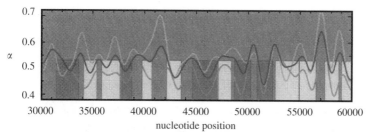

Fig. 4.42 Dependence of the exponent α on the nature (coding or noncoding) of the sequence analysed (after [2]).

For a fast Fourier transformation (FFT) analysis, we divide each sequence of L nucleotides into $K = [L/N]$ nonoverlapping subsequences of size $N = 512$ starting from the beginning and K nonoverlapping subsequences starting from the end of the sequence. For each subsequence we compute the Fourier transform and the power spectrum $S(f)$. Then we average $S(f)$ over the K subsequences of a given sequence, obtained from starting at one end, and K subsequences starting from the other end.

If a sequence has long-range power-law correlations, then

$$S(f) \sim f^{-\beta}, \tag{4.76}$$

and consequently a log–log plot of $S(f)$ versus f is a straight line with slope $-\beta$. The theoretical relation between α and β is given by

$$\alpha = (\beta + 1)/2. \tag{4.77}$$

It has been found that for the eukaryotic sequences for each of the mapping rules the average value of β is significantly smaller for coding sequences than for non-coding. The value of β is very close to zero for coding sequences, indicating almost no correlations in the region of 1/10 bp to 1/100 bp.

4.2.2 Word frequency analysis

In addition to the DNA walk, another interesting approach can be developed by considering DNA sequences as a text. Then one can apply various methods common in linguistics. In particular, we can define 'words' as short series of 'letters' and carry out a statistical analysis of word frequencies.

In n-tuple Zipf analysis, 'words' are defined as strings of n digits, and their normalized frequency of occurrence ω is measured for a given 'text'. It has been shown that such analysis leads to a power-law decay of word frequencies in the case of DNA texts [3].

4.2.3 Vector space techniques

As an alternative approach to the DNA walk, the symbol sequence corresponding to a DNA molecule can be regarded as a written text composed by using four letters. Naturally, we do not know the 'language' of the text; thus, when we are trying to get information about the sequence as a whole we are led to apply methods developed for analysing written (natural) texts of unknown origin.

Recently Damashek applied a vector-space technique to a number of texts written in different languages [4]. Using this approach he has been able to find correlations, for example, between texts of different origin, but written in the same language. The correlations of texts of similar content could also be manifested. Here two sequences are correlated if the scalar product of the two vectors associated with them has a value different from that which it would have for two uncorrelated sequences. In the first approximation his method is not sensitive to a class of encodings and can be used to identify the original language of encoded texts.

When applying the vector-space technique to DNA sequences, in a way we look at DNA as an encoded text written in an unknown language. Still, we expect to locate correlations between parts of the sequences due to similarities in their underlying structure.

The method

Let us move an n-character long 'window' along our document, symbol by symbol. An n-gram means a sequence (i.e., a string) of n given characters, that is, what can be 'seen in the window' at a particular position of the window. We denote each possible n-gram (a sequence of n consecutive characters) with an index i, and we count the number m_i of the occurrences of the ith n-gram in our text, for each possible i. The document can then be characterized by a 'document' vector x, whose components are the normalized occurrences as

$$x_i = \frac{m_i}{\sum_j m_j}. \tag{4.78}$$

The similarity of two texts with document vectors $\mathbf{x}$ and $\mathbf{y}$ can be measured by the dot product

$$S = \cos\theta = \frac{\sum_i x_i y_i}{(\sum_i x_i^2 \sum_i y_i^2)^{1/2}} \tag{4.79}$$

The maximum of this measure of similarity is 1, in the case of identical vectors, which occurs in practice only if the two documents are identical. The minimum of the dot product is zero in the case of orthogonal vectors, i.e., if there is no n-gram occurring in both documents. As discussed by Damashek [4], texts of the same language and topic result in a noticeably higher similarity than documents of different languages. The same language but different topics result in a dot product which is still significantly higher than the case of different languages, but usually smaller than the case of texts with related topics.

The procedure can be improved by introducing centroid vectors, which are the average of vectors taken from a given set of documents (e.g., the set of the documents in a given language). The centroid vectors contain the common average features of the document set (e.g., the grammatical words in a language). By subtracting the centroid vector from the document vectors, we can increase the sensitivity of the similarity measure. This method gives an effective technique called *Acquaintance* for sorting by language, topic and subtopic.

In the case of DNA sequences, our alphabet consists of four symbols, each representing one of the four bases. To demonstrate the method we chose five primary transcripts from the human haemoglobin sequence HUMHBB, NCBI-GenBank, Release 92.0 (Dec. 1995). Clay *et al.* [8] state that — according to their results — 'coding regions typically have higher GC levels than introns' of the same gene in human genome. (The GC level is the molar fraction of guanine + cytosine.) This statement

Table 4.2
Similarity among different regions of primary transcripts [5].
Note: E1–E5 denotes the concatenations of the exons of the five CDSs (coding regions) of HUMHBB, while I1–I5 denotes the concatenation of the corresponding non-coding sequences, i.e., heads, introns and tails. Our similarity measure, i.e., the dot product of the vectors representing the 'texts' to be compared ($n = 3$-grams were used), is symmetrical, and its range is between 0 and 1, where 0 means the lack of similarity, while 1 denotes maximal similarity. The similarity measure is significantly higher in the case of two coding or two non-coding sequences (0.94 ± 0.016 for exons, 0.92 ± 0.03 for non-coding 'texts') than in the case of a coding and a non-coding sequence (0.75 ± 0.08).

	E1	E2	E3	E4	E5	I1	I2	I3	I4	I5
E1	1.00	0.92	0.92	0.95	0.93	0.83	0.77	0.74	0.83	0.90
E2		1.00	0.97	0.93	0.95	0.73	0.66	0.62	0.73	0.85
E3			1.00	0.94	0.94	0.73	0.65	0.61	0.71	0.83
E4				1.00	0.95	0.78	0.71	0.67	0.79	0.87
E5					1.00	0.76	0.69	0.64	0.77	0.86
I1						1.00	0.92	0.90	0.94	0.93
I2							1.00	0.98	0.91	0.88
I3								1.00	0.90	0.86
I4									1.00	0.94
I5										1.00

predicts a difference in n-gram frequencies between coding and noncoding sequences, similar to the effect of phonological rules and orthographical traditions on document vectors.

We produced [5] two vectors from each primary transcript ($n = 3$): the first one representing the concatenation of the primary transcript's exons (E1–E5), the second one the concatenation of the head, the introns and the tail (I1–I5). Table 4.2 shows their dot products. The difference between the two groups is obvious. The similarity of two E-vectors is 0.94 ± 0.016, two I-vectors give 0.92 ± 0.03, while an E- and an I-vector produce 0.75 ± 0.08. The same result is obtained using the RATCRYG-sequence.

The results suggest the idea to use the 'moving box technique' for DNA-sequences, too. Figure 4.43 shows a primary transcript sequence of HUMHBB. The naive vector is calculated from the relative frequencies of the bases in the whole HUMHBB sequence ($n = 4$). A box of length $m = 100$ is moved along the primary transcript that contains a head, three exons, two introns and a tail. The pattern showed by the figure is characteristic of the HUMHBB-sequence: each exon is represented by a 'valley' in the figure, shifted by $m/2$ positions to the left (as explained in the case of natural languages); the first two of the three valleys overlap, and there is a fourth valley at about two-thirds of the primary transcript, before the last exon. This valley shows the existence of a sequence inside the second intron with statistical properties similar to those of the exons (or, at least, dissimilar to the statistical properties of other intron sequences).

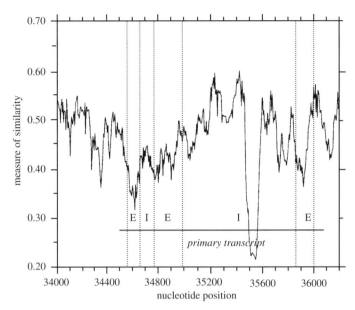

Fig. 4.43 The region of a primary transcript in the HUMHBB sequence (positions 34496,...,36087) that contains a head, three exons (E), two introns (I), and a tail. A box of length $m = 100$ was moved along the sequence, and at each step we plotted the dot product of the vector prepared from the actual content of the box with the naive (correlationless) vector, that we calculated from the base frequencies in the entire HUMHBB sequence. ($n = 4$ for both vectors.) Characteristic valleys can be seen where the box contains exon sequences, and another valley appears at about two-thirds of the primary transcript [5].

4.3 Analysis of brain electrical activity: the dimensional complexity of the electroencephalogram

M. Molnár

4.3.1 Neurophysiological basis of the electroencephalogram and event-related potentials

The excitable elements of the central nervous system (CNS) are nerve cells and glial cells. Action potentials and synaptic potentials can be recorded from the nervous system. Action potentials are several mV in size and their time course is rapid, i.e., they last only for a few ms. Synaptic (pre- and postsynaptic) potentials, on the other hand, are smaller (of the order of 10–500 uV) and their time course is longer (several hundreds of ms). According to the widely accepted view, the electroencephalogram (EEG) and also the event-related potentials (ERPs) elicited by sensory stimuli of different modalities are summations of synaptic potentials with very little – if any – contribution from action potentials.

The EEG

Traditionally, the spontaneous electrical activity produced by the brain is classified according to its frequency bands. The following characteristics are typical for the healthy adult. Alpha waves (8–12 Hz) occur typically in the awake, relaxed individual. The 13–20 Hz beta activity is typical of the alert state. Theta (4–7 Hz) and delta (0.5–3.5 Hz) waves are seen during sleep. Slow EEG waves usually have higher amplitudes (theta: 20–100 uV; delta: 20–200 uV) than faster rhythms (alpha: 20–60 uV; beta: 2–20 uV). Recently, the gamma band (30–50 Hz) has been payed increasingly more attention as it is supposedly related to various aspects of cognitive activity.

As a general rule, the two most important variables that define the EEG are (1) level of alertness (vigilance) and (2) age. The EEG is slower in sleep than in the awake state (except for the sleep stage that is characterized by low-amplitude fast EEG and rapid eye movements). The EEG recorded in children is slower than that seen in adults. Pathological processes alter the EEG, which in the case of tumours, vascular diseases, traumas, etc. is seen as an abnormal slowing. In epilepsy typical abnormal discharges are seen, characterizing the different types of this disease. The EEG is still the most important ancillary investigation for epilepsy.

Event-related potentials (ERPs)

ERPs, elicited by sensory events, are comprised of a series of waves having different polarities and latencies. These components are labelled according to their polarity (P or N) and latency; e.g., the N100 component is a negative wave with approximately 100 ms latency. Because of their small amplitude, averaging techniques are used to visualize ERPs by improving the 'signal-to-noise ratio'. The 'early' (also known as 'primary' or 'exogenous') components are usually defined as those that occur within the first 100 ms after the occurrence of the evoking stimulus and are determined by the physical characteristics of the evoking stimulus (e.g., louder clicks elicit larger potentials). The early ERP components are widely used in clinical practice for the investigation of the integrity of sensory pathways in the CNS. 'Late' components appear after the early ones and their characteristics are related to the psychological context in which the stimulus appears. In psychology and psychophysiology the term 'event-related potential' is widely used for potentials elicited by meaningful, task-related events [73].

4.3.2 Linear and nonlinear methods for the analysis of the EEG

Linear methods have been used in the past decades for this purpose with the assumption that the EEG is characterized by stochastic features, i.e., its features can be described by means, standard deviations, averages, etc. [74]. The amplitude and frequency characteristics can be described on this basis and scalp distributions of these features in a time-dependent manner can be visualized by specially developed techniques. The detailed description of these procedures and the rationale behind their use can be found in several handbooks [75, 76].

The stochastic (linear) approach, however, is not able to measure the nonlinear properties of the EEG, which may be a predominant characteristic feature [77, 78]. Previously, the focus of attention was mainly on linear systems analysis, but more recently new mathematical tools for the investigation of nonlinear dynamic systems have been introduced. The models of neuronal networks, often used for trying to simulate and analyse the behaviour of nerve cells and the CNS itself possess nonlinear wave-pulse functions. In a number of such networks, a simplification has been introduced: it was assumed that this nonlinear characteristic could be approximated by a linearized version. However, such simplification may lead one to ignore essential nonlinear properties of the system. Indeed, it is possible to account for the behaviour of neuronal networks assuming that these networks can be described by a number of coupled nonlinear differential equations as a function of time and space. In general, such a set of nonlinear differential equations may be considered to provide the basis for understanding the generation of all types of EEG activity [79]. A relevant aspect for this discussion is that all nonlinear dynamic systems with more than two degrees of freedom can display unpredictable behaviour over long timescales: in other words, they can display what mathematicians call chaotic behaviour [80]. A general problem with which one is faced is whether the ongoing EEG is a manifestation of a deterministic chaotic process having low degrees of freedom or a high-dimensional filtered noise [81]. The mathematical techniques with which to characterize the chaotic behaviour of nonlinear dynamic systems and to differentiate it from random activity are described in detail in a number of specialized publications. Here it is sufficient to state that the use of these mathematical tools opens promising new avenues for understanding the nature of complex patterns of EEG activity and merit further investigation.

Application of chaos theory to EEG analysis: the correlation dimension (D2)

Various procedures are known for assessing the nonlinear properties of the EEG, such as the Kolmogorov entropy , the Lyapunov exponent , or the correlation dimension [82, 83]. Of these, the latter seems to be favoured by most of the investigators [84].

The quantification of a chaotic system, such as the nervous system, can occur by calculating the correlation dimension (D2) of a sample of the data that the system generates. The dimension is the number of independent variables or degrees of freedom necessary for explaining the system's total behaviour or dynamics. The algorithm of Grassberger and Procaccia [85] has been used extensively to calculate the D2 of both biological and physical systems. Application of this mathematical tool to biological systems can determine whether the seemingly random behaviour of the intrinsic electrochemical processes are stochastic or are governed by the rules of deterministic chaos. What was thought to be higher-dimensional noise in many of the systems may turn out to be low-dimensional chaos. The term 'dimensional complexity' was coined by Pritchard and Duke [83], referring to the fact that the correlation dimension is one of those measures that can be applied to quantify the complexity of a system. Compared to that seen in normal individuals, decreased dimensional complexity was found in various pathological conditions such as in coma caused by Jakob–Creutzfeldt disease [84, 86] or in epilepsy [87, 88].

The correlation dimension: the D2 and PD2 algorithms

To determine a system's dimension its state space has to be constructed. The coordinates of this space are the degrees of freedom of the analysed system. It is possible to construct the state space from a single time series (such as the EEG), the structure of which is interpreted as involving a condensed image of the whole system generating the analysed signal. These time series can be continuously different or aperiodic. The dimensionality measures represent a 'pattern' that exists in the minimum phase space in which the time series is plotted (i.e., embedded). Thus the dimensional measures are sensitive in identifying these 'patterns' that can exist in aperiodic signals.

The correlation dimension (D2) of a time series is defined as $C(r, n) = r^{D2}$ where $C(r, n)$ is the cumulative of all rank-ordered vector-differences within a range (r) and n is the number of vector-differences. A reference vector $(nref)$ is made that begins at a specific point in the data and takes a specified number (m) of sequential time-steps in the data stream that are of a fixed length (tau); each value encountered in the time-steps is used as one coordinate of the m-dimensional vector. A different vector is then made by moving to a new starting point, for example, to the next point in the time-series, and then using the same number of tau-steps. Then another vector is made by starting at the third point in the series, and so on for all of the points in the data series. All possible vector-differences for every possible $nref$ made with a given embedding dimension (i.e., number of tau-steps, m) are then rank-ordered and a log $C(r, n)$ versus log r plot is made. The slope of the linear region in this plot is then measured; this linear region reflects the range of r over which the model $C(r, n) = r^{D2}$, or D2 $= \log C(r, n)/\log r$, is valid. The value of m is incremented and the corresponding slope noted, thus yielding slope and m pairs. The values of m are selected to span the size of the expected D2 value (that is, m ranges from 1 to 2D2 + 1). The number of embedding dimensions is relevant up to the point where its increment is no longer associated with an increase in slope (i.e., it converges). D2 then is the slope of the linear region at the convergent values of m.

Mathematically stationary data are presumed in the above application, a presumption which is rarely tenable for biological data, as the generator is constantly changing. The algorithm found be the most accurate in estimating the dimension from limited data is the 'Point Correlation Dimension' (PD2) which was developed by Skinner et al. [89]. This algorithm does not presume mathematically stationary data, as does the D2 algorithm of Grassberger and Procaccia. Rather the PD2 tracks the transient nonstationary changes that occur when the generator changes state, as often happens in biological systems (e.g., during behavioural arousal, epileptic seizure, or heart attack).

The 'PD2' estimate of the correlation dimension was found to reduce the variance of the estimates compared to the 'pointwise' D2 method. The PD2 does not use all possible vector-differences, like the Grassberger and Procaccia algorithm, nor all vector-differences with respect to a fixed $nref$. Rather, it rejects those $nref$ vector-differences for which linear scaling and smooth convergence cannot be found. Accepting every data-point as an $nref$ means erroneously including those vectors for which the relationship $C(r, n) = r^{D2}$ does not hold. In other words, the dimension cannot be estimated at some $nref$ because the data points, being finite, are not distributed on the m-dimensional 'strange-attractor' in a manner suitable for estimating the correlation dimension starting at that particular $nref$. The model for the PD2 is $C(r, n, nref^*) = r^{D2}$, where $nref^*$ passes two criteria: (a) linear scaling in the $\log C(r, n, nref)$ versus $\log r$ plots and (b) convergence of slope versus m.

The value of tau is irrelevant if the number of points in the time-series is infinite, a condition which is never approached for biological data. A conventional way of determining tau to use is to calculate the first zero-crossing of the autocorrelation function of the data time-series (i.e., approximately one quarter cycle of the dominant frequency). One should be cautious about new tau requirements when nonstationary changes arise in finite data; the autocorrelation function and tau selection should be evaluated for each subepoch.

The precision of the PD2 algorithm was tested for estimating the D2 of several linked time-series, each of which was generated by a mathematical function for which the dimension is different and is known: the sine (D2 = 1.00), the Lorenz (D2 = 2.06), and the Henon (D2 = 1.26) functions [90]. The 'sampling rate' of each series was made so that tau= 1 could be used (the Henon series naturally has tau= 1). It was found that the accuracy of the PD2 (subepoch mean) for each of these finite data samples was superior to that of the classical D2 algorithms [85] or that of the 'pointwise' scaling dimension D2i. Thus the use of the PD2 algorithm to assess biological data is: (a) appropriate, as it meets all data requirements (i.e., it does not require mathematically stationary data and thus can analyse small subepochs or 'points'), (b) accurate, even in its estimate of D2 using relatively small epochs of data, and (c) timely, as recent dimensional assessment has been proposed to reveal deterministic processes that may underline poorly understood phenomena, such as cognition and sudden cardiac death.

It is to be noted that for dimensional analysis the low number of data points (as in the present study) is not optimal. In biological systems, however, it is not possible to acquire the few millions of data points that would be desirable to satisfy theoretical requirements. A higher digitization rate, although resulting in more data points per epoch, would introduce problems of oversampling. For this reason, it seems justifiable to use the term 'effective point correlation dimension' (PD2$_{eff}$) in cases when particularly short data epochs are used, i.e., in ERP analysis.

4.3.3 Examples of the application of PD2 to EEG and ERP analysis

Experimental data are shown below to illustrate the efficacy of nonlinear data analysis – and particularly that of PD2 – in various fields of electrophysiology. Part of the experiments were performed in the Department of Neurology, Section of Neurophysiology, Baylor College of Medicine, Houston, Texas; others in the Institute of Psychology of the Hungarian Academy of Sciences, Department of Psychophysiology, Budapest.

Pharmacological studies

In recent years the potential advantages offered by nonlinear methods have been utilized in the testing of different drugs [91]. In the following a typical example is shown in which the use of the PD2 method was tested.

The experiments were performed in mice. The EEG was recorded by a screw electrode placed in the skull bone (vertex) with occipital reference by Grass amplifiers (frequency band: DC–100 Hz). The EEG was digitized by a Compaq computer at 200 Hz.

In Fig. 4.44. the effect of the anaesthetic Ketamine (15 mg/kg) is shown. The PD2, calculated from the corresponding EEG epoch, is shown as a dot-diagram. The high-frequency components of the EEG were wiped out by the drug which was dominated by theta–delta activity and the PD2 calculated from the EEG was significantly ($p <$ 0.001) lower than that of the control. A similar result was found for atropine, inducing high-amplitude slow activity with the reduction of fast waves. These findings are to be expected since the anaesthetic drug (and the cholinergic blocker atropine) causes a 'simpler' activity characterized by the loss of fast, 'irregular' waves and consequently lower dimensional complexity. It is to be emphasized, however, that the frequency and the dimensional characteristics of the EEG are independent of each other [92].

Sleep studies

Low correlation dimension was reported both in human and in animal studies for the EEG recorded in 'slow wave sleep' (SWS), i.e., that period of sleep which is characterized by slow (theta and delta) waves in the EEG [92, 93]. The result of a human study is shown in Fig. 4.45. The EEG was recorded from the vertex in the quiet, awake state (QUIET) and during increasingly deepening sleep (from SWS1 through SWS2 to SWS3). The EEG was recorded also in REM ('rapid eye movements': a sleep stage characterized by low-voltage desynchronized EEG and total loss of muscle tone) sleep. The frequency power spectra and the histograms of the PD2 values calculated from each epoch are shown on the right-hand side. Consecutively lower PD2 values were found in deepening sleep stages. In REM sleep, PD2 was found to increase again.

The interpretation of the above findings is somewhat similar to that offered for the drug studies; lower PD2 can be expected to be found for an EEG that is dominated

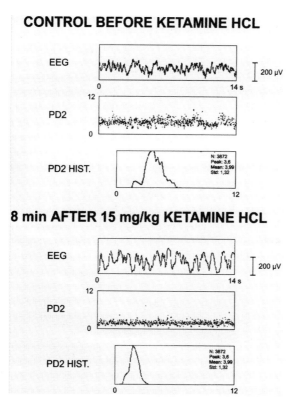

CONTROL BEFORE KETAMINE HCL

8 min AFTER 15 mg/kg KETAMINE HCL

Fig. 4.44 The effect of Ketamine HCl on the EEG and on the PD2, which latter is indicated below the EEG traces as a dot-diagram. Control is in the upper part, and the effect is in the lower part of the figure. Histograms of the PD2 values are also shown (PD2 HIST).

by synchronized, oscillatory waves, as in the case of SWS. Subcortical (probably thalamic) generators are considered responsible for this kind of activity, driving the cortex via the cortico-thalamo-cortical loops. Contrary to this condition, during the awake state and during REM sleep when the EEG is characterized by fast ('desyn-chronized') rhythms, the PD2 was found to be high, corresponding to the increasing number of interacting active elements, i.e., neurons or neuronal networks. In other words, higher PD2 seen in the alert state and in REM corresponding to the increased degrees of freedom (i.e., the number of independently active neurons) reflected in desynchronized activity [94]. Nonlinear methods were claimed to be more sensitive than linear ones in discriminating between different stages of sleep [95].

4.3.4 Dimensional analysis of ERPs

Human studies

In the past decades ERPs have proven to be sensitive indices of various aspects of sensory information processing in the nervous system. Thus, for example, the P3

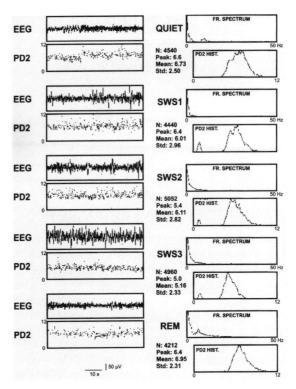

Fig. 4.45 The effect of different sleep stages on the PD2. The EEG and corresponding PD2 values are on the left; the frequency spectra (FR.SPECTRUM) and PD2 histograms (PD2 HIST) for each sleep stage are shown on the right. Below the individual definition of sleep stages (QUIET, SWS1-SWS3, REM) the actual values (Peak, Mean, Standard deviation [Std] and number of accepted data points [N] for the calculation) can be seen.

wave is an ERP component elicited by randomly occurring rare odd-ball stimuli, but it will appear only if the subject is instructed to respond to these stimuli in some way. Although the experimental circumstances in which these 'cognitive' components will appear are rather clear, less is known about the psychophysiological mechanisms they represent. The application of recently developed nonlinear methods may help to extend the possibilities of interpretation, not only for the spontaneous activity of the CNS but also for ERPs. Only in a few studies were the effects of cognitive 'effort' investigated on the D2 calculated from the EEG in which an increase of the D2 was shown as a consequence of cognitive 'load' [96]. In one study, based on the analysis of single EEG epochs recorded from Pz in an auditory odd-ball experiment it was found that the value of D2 decreased, which was associated with the occurrence of the target stimulus [97]. Some related experiments are described below.

Thirteen normal healthy adults were used as subjects. The auditory stimuli were 50 ms long tones of 80 dB SPL intensity delivered binaurally through earphones at a rate of 1/s. The EEG was recorded by a multichannel amplifier system in the midline from the frontal, central, and parietal area with bandpass filtering of 0.1–150 Hz. Stimulus delivery and digitization of the EEG were both carried out by IBM-compatible computers. Background ('standard') auditory stimuli of 1000 Hz frequency were randomly (probability: 10%) interrupted by 'deviant' stimuli of 1100 Hz. In the first ('ignore') part of the experiment the subjects were instructed not to pay any attention to the stimuli at all. In the second ('count') part of the experiment the subjects were asked to count the occurrence of higher frequency 'target' stimuli. When the deviant stimuli became targets, these evoked the P3 ERP component accompanied by a clear drop in the corresponding $PD2_{eff}$ [98]. The $PD2_{eff}$ decrease was statistically significant ($p < 0.0001$), showed regional specificity and the pattern depended upon task difficulty [99].

Animal studies

The experiments were carried out in 8 chronically implanted cats. The animals were treated in a humane manner as outlined in the Helsinki Accords for Animal Welfare. In Nembutal anaesthesia, electrodes (0.23 mm stainless steel) were placed on the dura above the primary auditory cortex, association cortical areas and dorsal hippocampus. A 'vertex' screw electrode was placed in the skull 1 cm anterior to the bregma in the midline. Additional electrodes were placed in the periorbital fat tissue (for recording the electrooculogram, EOG) and in the superficial neck muscles. Grounding electrode was placed in the occipital bone and the reference electrode in the bone covering the nasal sinuses in the midline. The electrodes were led into a miniature Winchester socket embedded in dental cement on the top of the skull. The electrical activity was led into high input impedance amplifiers through a light, low-noise cable which did not interfere with the movement of the animals, the behaviour of which was monitored by a closed-loop TV system. The EEG was digitized by Neuroscan software at 1000 Hz and was stored on hard disk for off-line analysis.

Auditory stimuli (80 dB SPL) were delivered through a bone conductor, for which a place was formed in the dental acrylic, to maintain stable acoustic input. Classical aversive conditioning was used. First, the animals were habituated to 3 ms long tones of 4 and 2 kHz frequency presented continuously at 1/s. At random intervals (probability: 5%) the 4 kHz tones were interrupted by a 2 kHz tone. During conditioning, 800 ms after the 2 kHz tone, an unavoidable electric shock was delivered through the muscle electrodes. A 2 s long trial consisted of a 4 kHz 'background' tone followed by a 'signal' tone. The intertrial interval was 2–8 min. Daily sessions consisted of 50 trials.

ERPs elicited by the 4kHz tones and those by 2 kHz tones were averaged separately. PD2 analysis was performed on all digitized epochs in all animals, recorded in all electrode positions. Following PD2 analysis, conventional averaging was performed both on the EEG epochs, yielding the ERPs, and also on the determined dimensional data for each digitized data epoch. Grand averages of the data obtained in the 8

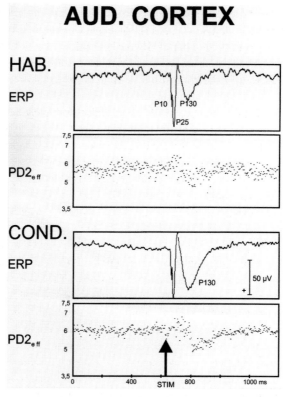

Fig. 4.46 Grand means of auditory event-related potentials and corresponding PD2$_{eff}$ values recorded from 8 cats after habituation (HAB.) and as a result of conditioning (COND.). The timing of the auditory stimulus is indicated by the arrow. For clarity the P10 and P25 components are shown only for the habituated potential.

animals were then made. In this study, ERPs evoked by the 'signal' tones in the primary auditory cortex, and the PD2$_{eff}$ changes associated with these responses are shown. The results using the value of tau shown in this study was 10, referring to a fundamental frequency of 25 Hz. Other tau values (25 and 18) were also tested and were found to cause slightly different PD2 change patterns. However, the nature of the PD2$_{eff}$ change was similar in all cases. During the calculation of the PD2$_{eff}$, a 'moving window' of tau × embedding dimension (e.g., in this case, 10 × 12 = 120 ms) latency range is inherent in the procedure. Since the calculation is performed in a time-dependent manner, a correction for this uncertainty can be achieved if the end-section of the epochs are disregarded where because of lacking data points no additional vectors can be made [98], which was taken into account in the present analysis also. Student's t-test was used for statistical evaluation.

Grand averages of auditory ERPs elicited by the randomly occurring 2 kHz stimulus and associated PD2$_{eff}$ changes seen in 8 cats are shown in Fig. 4.46, observed in the habituated and conditioned animals. The PD2$_{eff}$ values were measured as baseline 400 ms before the 2 kHz stimulus, from stimulus onset till the peak of the P130 wave, and from this peak for 200 ms. In the habituated condition the P10, P25 and P130 ERP components were seen. These were accompanied by a slight increase (from 5.67 to 5.76; not significant) of the PD2$_{eff}$ from baseline followed by a decrease (to 5.28; $p < 0.001$).

As a result of conditioning the amplitude of the P130 component of the ERP significantly ($p < 0.001$) increased (from 50 uV to 77 uV). The pattern of the PD2$_{eff}$ change became more robust: both the initial increase and the decrease following it were of higher magnitude (baseline: 5.93; initial increase: 6.09 [$p < 0.001$]; later decrease: 5.40 [$p < 0.0001$]). The baseline PD2$_{eff}$ was significantly ($p < 0.0001$) higher during the conditioning than during habituation.

It could be expected that a dimensional increase would accompany the occurrence of the P3 component, which latter indicates a sensory discrimination and decision making in the processing of the evoking stimulus. Dimensional increase accompanying mental effort was found in the spontaneous EEG [96, 100, 101]. Instead, a conspicuous decrease was found both in animals and in humans, although the initial increase was conspicuous in the animal studies. On the basis of animal studies carried out in the olfactory bulb of the rabbit several authors suggested that the presentation of a learned stimulus would switch the system's state towards a limit cycle attractor having low dimension, while novel stimuli increase the dimension [102, 103]. Our finding is congruent with literature data according to which such a dimensional decrease accompanied execution of mental tasks [104] and memory load [105]. Dimensionality becomes reduced over those brain areas which become actively involved in the task currently executed [82].

One of the most well-known hypotheses on the psychological mechanism(s) that the P3 represent suggests that it corresponds to a 'context updating' process [106]. This implies that subjects have a 'model' of the experimental situation and if a stimulus causes that model to change, it needs to be updated, which process is reflected by the occurrence of the P3. According to a more recent hypothesis the odd-ball stimulus causes 'context closure', and this event corresponds to the occurrence of the P3. This theory [107] suggests that the occurrence of the P3 ERP component signals the end of the decision-making process related to the actual task. From this standpoint, the occurrence of the target stimulus, eliciting the P3 component, is regarded as such, which causes a temporary relaxation which results in the decrease of dimension and thus is a manifestation of an inhibitory process. Equally potent and not exclusive of this hypothesis, however, is the possibility that the dimensional decrease reflects increased cooperation between neuronal networks – taken in the broad, Hebbian sense of the word – that are responsible for the analysis of the meaningful, task-related sensory event. On the other hand, the increased baseline PD2 during conditioning may indicate a general arousal effect, supposedly caused by the aversive nature of the experimental situation.

No P3 was expected to occur on the auditory cortex in the cat experiments, since this component, the origin and neural mechanisms of which are still obscure, is known to be present with largest amplitude on the vertex and parietal areas [108]. However, our finding that in the latency range of the P130 component a significant change (i.e., increasing amplitude of this wave) occurred as a result of conditioning is a new observation, since according to the generally accepted view no such ERP change is expected in primary sensory areas in such circumstances [109, 110]. Thus it appears that the results of both the dimensional analyses and the ERP observations support our previous findings [111] that significant changes can be observed on the primary auditory cortex during the development of classical aversive conditioning, which phenomenon was related to a 'learned aversive arousal' process.

The obvious difference between the results of the animal and human studies (i.e., the presence of an initial dimensional increase elicited by the stimulus) may be explained by taking into account the fact that in the case of scalp recordings volume-conducted activity is recorded only, while in the case of the animal studies the analysed signals reflect activity produced by the generator(s) picked up by closely placed electrode(s). The presence of the dimensional decrease observed in both cases implies that for its interpretation the restrictions posed by the analysis of volume-conducted activity (i.e., undefined generators, etc.) may not necessarily apply. However, in both cases the PD2 decrease occurred in the long latency range when volume conduction is likely to play an increased role.

4.3.5 Clinical applications

Epilepsy

Just as in the case of the occurrence of lethal arrhythmias [89], it has been proposed that application of nonlinear measures to the EEG may prove to be a useful tool for the prediction of the occurrence of epileptic seizures [112, 113]. Localization of epileptic focus was the other ultimate goal of these efforts [88]. An example is shown below for this latter possibility.

A 14-year-old boy was only partly successfully treated for epileptic seizures. The seizures started as partial ones but generalized within seconds, causing loss of consciousness. The patient's intellectual development deteriorated and he showed behavioural problems of increasing severity. In the upper part of Fig. 4.47 multichannel epicortical (1–15) recordings of the EEG are shown during a generalized seizure elicited by electrical stimulation during neurosurgery. Below the EEG traces the PD2 values calculated from each recording are shown as dot-diagrams. The values of PD2 scatter widely during the development of the seizure. In the lower part, for clear visualization only two (1 and 6) channels are shown. At the location of electrode 6, which proved to be the site of origin of the seizures as verified by direct electrocorticography performed by stimulation, the neurosurgeon removed an extremely small developmental malformation. As can be seen, when compared to other locations such as at

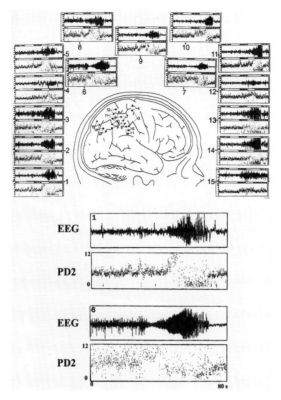

Fig. 4.47 *Upper part*: 15 channels of EEG and corresponding PD2 recorded during a generalized seizure. Positions of the epicortical electrodes are indicated on the graph in the middle. *Lower part*: enlarged traces for electrodes 1 and 6.

electrode 1, at this cortical site the standard deviation of the PD2 before the seizure was conspicuously higher. This feature of the EEG can probably be regarded as a pathological sign characterizing the abnormal functional state of the epileptogenic area.

Stroke

The hypothesis behind these studies was that by the use of this approach additional information can be obtained concerning the functional state of different cortical areas. Mapping of the scalp distribution of the PD2, calculated from the spontaneous EEG, was performed in patients who had previously suffered stroke (partial occlusion of the middle cerebral artery). These findings were correlated with the results of conventional clinical electrophysiological methods.

The EEG was recorded for 2 min in the 'eyes closed', 'eyes open' and video viewing conditions by 32 electrodes using Neuroscan amplifiers. 80 s artefact-free epochs were subjected to dimensional analysis, before which 0.5 Hz off-line high-pass filtering was carried out. The PD2 was calculated from every EEG epoch recorded at all electrode locations. After the PD2 calculation, an overall average of the PD2 values seen at all electrode locations were made, of which a grand average was then calculated for all subjects characterizing the condition in which it was observed. Z-score maps of the PD2 scalp-distributions were then calculated. Differences from these average values were plotted in units of standard deviation. Grand averages of the PD2 observed in healthy, age-matched subjects were used as control.

In the healthy controls the scalp distribution of the PD2 values was symmetrical in all recording conditions. In the case of the patients, however, the distribution of PD2 was markedly asymmetrical, with low values corresponding to the side of the infarct. The extension of the low-dimensional area depended on the recording condition, being larger in the 'eyes closed' than in the 'eyes open' or 'video viewing' condition. Analysis of frequency power spectra revealed an only partly overlapping slow focus of activity. In one case of subcortical infarct the only sign of pathology on the same side verified by conventional methods was the relative reduction of the fast (gamma band) activity and reduced amplitude of the P3 ERP component on the same side [114].

It is not surprising that conspicuously lower than normal dimensions were reported in earlier investigations, when EEG recorded in cases with severe pathology like coma or epilepsy were analysed. In these cases the EEG itself is informative of the abnormal clinical condition and the pathological process involving the whole brain. In our present cases, however, the patients had focal abnormalities caused by stroke. The fact that we found higher dimension values in all electrode sites during the 'eyes open', than the 'eyes closed' condition does not seem surprising, since the activation caused by opening of the eyes can be expected to cause such an effect and agrees with the literature data. It has yet to be seen how the changes of the low-dimensional cortical areas caused by different activation techniques and visualized by correlation dimension mapping can be correlated with the severity of the clinical condition and the long-term outcome of the patient. Our conclusion of these studies was that methods developed on the basis of chaos theory may not only be complementary to those of the traditional electrophysiological ones but can also be superior to them in the assessment of the functional condition of different cortical areas. At least in some types of neurological diseases the information that the nonlinear methods yield may in the future play an important role in diagnostic, and perhaps in therapeutic, issues as well.

4.3.6 Concluding remarks

Application of chaos theory to biological systems analysis presents problems and promises alike. The major source of problems are both technical (use of different algorithms, different recording conditions, etc.) and theoretical (interpretation of the results). Perhaps one of the hardest obstacles to be overcome is caused by the fact that the consequences and advantages of the nonlinear approach have to be translated for the life sciences from other fields, since the origins of this methodology can be mostly traced back to mathematical and physical theories and applications. It would be probably a rather naive attitude to suppose that the application of nonlinear techniques will solve the multitude of problems posed by the analysis of different features of brain electrical activity, especially those features that are commonly referred to as 'higher brain functions'. It is probably not by chance, however, that ideas such as 'self-organization', possibly related to elementary and crucially important modes of function of the CNS, received special attention from applicants of these new techniques and that these findings were incorporated into more general theories concerning basic functions of the nervous system [77]. Thus, the prejudiced investigator who claims that nonlinear methods guarantee salvation can hardly be excused except to say that nonlinear characteristics do appear to dominate the electrophysiological features of brain activity and thus their disciplined use is to be welcome. Indeed, methodological issues seem to be less frequent in recent publications and it seems justifiable to expect that this new approach will probably allow a fresh new look at old and currently arising challenges concerning the use of electrophysiological methods and the interpretation of data in the investigation of the CNS.

References

[1] C.-K. Peng, S. V. Buldyrev, A. L. Goldberger, S. Havlin, F. Sciortino, M. Simons, and H. E. Stanley. Long-range correlations in nucleotide sequences. *Nature* 356: 168–170, 1992.

[2] H. E. Stanley, S. V. Buldyrev, A. L. Goldberger, S. Havlin, S. M. Ossadnik, C.-K. Peng, and M. Simons. Fractal landscapes in biological systems. *Fractals* 1: 283–301, 1993.

[3] A. Czirók, R. N. Mantegna, S. Havlin, and H. E. Stanley. Correlations in binary sequences and a generalized Zipf analysis. *Phys. Rev. E* 52: 446–452, 1995.

[4] M. Damashek. Gauging Similarity with *n*-Grams. Language-Independent Categorization of Text. *Science* 267: 843-848, 1995.

[5] T. Biró, A. Czirók, T. Vicsek, and Á. Major. Applications of vector space techniques to DNA. *Fractals* 6: 205–210, 1998.

[6] C. Allison, and C. Hughes. Bacterial swarming. an example of procaryotic differentiation and multicellular behaviour. *Sci. Prog.* 75: 403–422, 1991.

[7] D. G. Aronson, and H. F. Weinberger. Multidimensional nonlinear diffusion arising in population genetics. *Adv. Math.* 30: 33, 1978.

[8] A.-L. Barabási , and H. E. Stanley. *Fractal concepts in surface growth*. Cambridge University Press, Cambridge, 1995.

[9] E. Ben-Jacob, H. Brand, G. Dee, L. Kramer, and J. S. Langer. Pattern propagation in nonlinear dissipative systems. *Physica D* 14: 348–364, 1985.

[10] E. Ben-Jacob, G. Deutscher, P. Garik, N. Goldenfeld, and Y. Lereay. Formation of a dense branching morphology in interfacial growth. *Phys. Rev. Lett.* 57: 1903, 1986.

[11] E. Ben-Jacob, I. Cohen, A. Cziók, T. Vicsek, and D. L. Gutnick. Chemomodulation of cellular movement, collective formation of vortices by swarming bacteria and colonial development. *Physica A* 238: 181–197, 1997.

[12] E. Ben-Jacob, I. Cohen, I. Golding, and Y. Kozlovsky. Modeling branching and chiral colonial patterning of lubricating bacteria. *(preprint, cond-mat/9903382)*, 1999.

[13] E. Ben-Jacob, I. Cohen, O. Shochet, A. Czirók, and T. Vicsek. Cooperative formation of chiral patterns during growth of bacterial colonies. *Phys. Rev. Lett.* 75: 2899–2902, 1995.

[14] E. Ben-Jacob, H. Shmueli, O. Shochet, and A. Tenenbaum. Adaptive self-organization during growth of bacterial colonies. *Physica A* 187: 378–424, 1992.

[15] E. Ben-Jacob, O. Shochet, I. Cohen, A. Tenenbaum, A. Czirók, and T. Vicsek. Cooperative strategies in formation of complex bacterial patterns. *Fractals* 3: 849–868, 1995.

[16] E. Ben-Jacob, O. Shochet, A. Tenenbaum, I. Cohen, A. Czirók, and T. Vicsek. Communication, regulation and control during complex patterning of bacterial colonies. *Fractals* 2: 14–44, 1994.

[17] E. Ben-Jacob, O. Shochet, A. Tenenbaum, I. Cohen, A. Czirók, and T. Vicsek. Generic modelling of cooperative growth patterns in bacterial colonies. *Nature* 368: 46–49, 1994.

[18] E. Ben-Jacob, A. Tenenbaum, O. Shochet, and O. Avidan. Holotransformations of bacterial colonies and genome cybernetics. *Physica A* 202: 1–47, 1994.

[19] A. Buka, J. Kertész, and T. Vicsek. Transitions of viscous fingering patterns in nematic liquid crystals. *Nature* 323: 424, 1986.

[20] I. Cohen, A. Czirók, and E. Ben-Jacob. Chemotactic-based adaptive self-organization during colonial development. *Physica A* 233: 678–698, 1996.

[21] Z. Csahók, K. Honda, and T. Vicsek. Dynamics of surface roughening in disordered media. *J. Phys. A* 26: L171–L178, 1993.

[22] A. Czirók, E. Ben-Jacob, I. Cohen, and T. Vicsek. Formation of complex bacterial patterns via self-generated vortices. *Physical Review E* 54: 1791–1801, 1996.

[23] A. Czirók, Z. Csahók, and T. Vicsek. unpublished results.

[24] M. M. Dombach, S. K. Leung, R. E. Cahn, G. G. Cocks, and M. L. Shuler. Computer model for glucose-limited growth of a single cell of *escherichia coli* b/r-a. *Biotechnol. Bioeng.* 26: 203–216, 1984.

[25] M. Eden. A two-dimensional growth process. In *Biology and Problems of Health*, Vol. 4. University of California Press, Berkeley, 1961, pp. 223–39.

[26] Essam, Guttmann, and De'Bell. On two-dimensional directed percolation. *J. Phys. A* 21: 3815, 1988.

[27] W. W Ford. Studies on aerobic spore-bearing non-pathogenic bacteria. Part ii: miscellaneous cultures. *J. Bacteriology* 1: 518–526, 1916.

[28] H. Fujikawa and M. Matsushita. Fractal growth of *bacillus subtilis* on agar plates. *J. Phys. Soc. Jpn.* 58: 3875–3878, 1989.

[29] H. Fujikawa and M. Matsushita. Bacterial fractal growth in the concentration field of nutrient. *J. Phys. Soc. Jpn.* 60: 88–94, 1991.

[30] T. Halpin-Healy and Y.-C. Zhang. Kinetic roughening phenomena, stochastic growth, directed polymers and all that. *Phys. Rep.* 254: 218–405, 1995.

[31] J. W. Jeong, J. Snay, and M. M. Ataai. A mathematical model for examining growth and sporulation processes of *bacillus subtilis*. *Biotechnol. Bioeng.* 35: 160–184, 1990.

[32] A. Joshi and B. O. Palsson. *Escherichia coli* growth dynamics: a three-pool biochemically based description. *Biotechnol. Bioeng.* 31: 102–116, 1988.

[33] R. Jullien and R. Botet. Scaling properties of the surface of the eden model in d=2,3,4. *J. Phys. A* 18: 2279–2287, 1985.

[34] M. Kardar, G. Parisi, and Y.-C. Zhang. Dynamic scaling of growing interfaces. *Phys. Rev. Lett.* 56: 889–892, 1986.

[35] A. L. Koch. Biomass growth rate during the prokaryote cell cycle. *Crit. Rev. Microbiol.* 19: 17–42, 1993.

[36] J.-U. Kreft, G. Booth, and J. W. T. Wimpenny. Bacsim, a simulator for individual-based modelling of bacterial colony growth. *Microbiolgy* 144: 3275–3287, 1998.

[37] J. S. Langer. Studies in the theory of interfacial instability II. Moving symmetric model. *Acta Metall.* 25: 1121, 1977.

[38] J. S. Langer. Instabilities and pattern formation in crystal growth. *Rev. of Mod. Phys.* 52: 1–27, 1980.

[39] H. Leschhorn. Anisotropic interface depinning: numerical results. *Phys. Rev. E* 54: 1313, 1996.

[40] M. Matsushita, M. Sano, Y. Hayakawa, H. Honjo, and Y. Sawada. Fractal structures in zinc metal leaves grown by electrodeposition *Phys. Rev. Lett.* 53: 286, 1984.

[41] M. Matsushita, J. Wakita, H. Itoh, I. Ràfols, T. Matsuyama, H. Sakaguchi, and M. Mimura. Interface growth and pattern formation in bacterial colonies. *Physica A* 249: 517–524, 1998.

[42] M. Matsushita and H. Fujikawa. Diffusion-limited growth in bacterial colony formation. *Physica A* 168: 498–506, 1990.

[43] T. Matsuyama, R. M. Harshey, and M. Matsushita. Self-similar colony morphogenesis by bacteria as the experimental model of fractal growth by a cell population. *Fractals* 1: 302–311, 1993.

[44] T. Matsuyama and M. Matsushita. Self-similar colony morphogenesis by gram-negative rods as the experimental model of fractal growth by a cell population. *Appl. Env. Microbiology* 58: 1227–1232, 1992.

[45] T. Matsuyama and M. Matsushita. Fractal morphogenesis by a bacterial cell population. *Critical Reviews in Microbiology* 19: 117–135, 1993.

[46] T. Matsuyama and M. Matsushita. Comments on 'classification and genetic characterization of pattern-forming *bacilli*'. *Molecular Microbiology* 31: 1278–1279, 1999.

[47] P. Meakin. Accretion processes with linear particle trajectories. *J. Colloid Interface Sci.* 105: 240, 1985.

[48] W. W. Mullins and R. F. Sekerka. Morphological stability of a particle growing by diffusion or heat flow. *J. Appl. Phys.* 34: 323, 1963.

[49] J. D. Murray. *Mathematical Biology.* Springer–Verlag, Berlin, 1989.

[50] O.M. Neijssel, M.J.T. de Mattos, and D.W. Tempest. *Growth yield and energy distribution*, (2nd edn). American Society for Microbiology, Washington, DC, 1996, pp 1683–1692.

[51] M. Ohgiwari, M. Matsushita, and T. Matsuyama. Morphological changes in growth phenomena of bacterial colony paterns. *J. Phys. Soc. Jpn.* 61: 816–822, 1992.

[52] L. M. Prescott, J. P. Harley, and D. A. Klein. *Microbiology, (3rd edn).* Wm. C. Brown, Dubuque, IA, USA, 1996.

[53] G. Radnóczy, T. Vicsek, L. M. Sander, and D. Grier. Growth of fractal crystals in amorphous $GeSe_2$ films. *Phys. Rev. A* 35: 4012, 1987.

[54] O. Rauprich, M. Matsushita, C. J. Weijer, F. Siegert, S. E. Esipov, and J. A. Shapiro. Periodic phenomena in *proteus mirabilis* swarm colony development. *Journal of Bacteriology* 178: 6525–6538, 1996.

[55] R. Rudner, O. Martsinkevich, W. Leung, and E. D. Jarvis. Classification and genetic characterization of pattern-forming *bacilli*. *Molecular Microbiology* 27: 687–703, 1998.

[56] R. F. Sekerka. A time-dependent theory of stability of a planar interface during dilute binary alloy solidification. *J. Phys. Chem. Solids* 28: 691, 1967.

[57] J. A. Shapiro. Bacteria as multicellular organisms. *Sci. Am.* 258: 62–69, 1988.

[58] J. A. Shapiro. The significances of bacterial colony patterns. *BioEssays* 17: 597–607, 1995.

[59] J. A. Shapiro and M. Dworkin, editors. *Bacteria as multicellular organisms.* Oxford University Press, Oxford, 1997.

[60] R. N. Smith and Clark F. E. Motile colonies of *bacillus alvei* and other bacteria. *J. Bacteriology* 35: 59–60, 1938.

[61] M. Tcherpakov, E. Ben-Jacob, and D. L. Gutnick. *Paenibacillus dendritiformis* sp. nov., proposal for a new pattern-forming species and its localization within a phylogenetic cluster. *Int. J. Syst. Bacteriol.* 49: 239–245, 1999.

[62] D. W. Tempest and O. M. Neijssel. The status of y_{ATP} and maintenance energy as biologically interpretable phenomena. *Annu. Rev. Microbiol.* 38: 459–486, 1984.

[63] T. Vicsek, M. Cserzǫ, and V. K. Horváth. Self-affine growth of bacterial colonies. *Physica A* 167: 315–321, 1990.

[64] T. Vicsek, A. Czirók, O. Shochet, and E. Ben-Jacob. Self-affine roughening of bacterial colony surfaces. In *Spatio-temporal patterns* (ed. P. Palffy-Muhoray and P. Cladis). Santa-Fe Inst. Proc. Vol XXI. Addison Wesley, New York, 1995.

[65] T. Vicsek. *Fractal growth phenomena*, (2nd edn). World Scientific Publishing Co., Singapore, 1992.

[66] R. F. Voss. Multiparticle fractal aggregation. *J. Stat. Phys.* 36: 861–872, 1984.

[67] J. Wakita, H. Itoh, T. Matsuyama, and M. Matsushita. Self-affinity for growing interface of bacterial colonies. *J. Phys. Soc. Jpn.* 66: 67–72, 1997.

[68] J. Wakita, K. Komatsu, A. Nakahara, T. Matsuyama, and M. Matsushita. Experimental investigation on the validity of population dynamics approach to bacterial colony formation. *J. Phys. Soc. Jpn.* 63: 1205–1211, 1994.

[69] X. R. Wang, Y. Shapir, and M. Rubinstein. Analysis of multiscaling in diffusion-limited aggregation: a kinetic renormalization approach. *Phys. Rev. A* 39: 5974–5984, 1989.

[70] J. W. T. Wimpenny and R. Colasanti. A unifying hypothesis for the structure of microbial biofilms based on cellular automaton models. *FEMS Microbiology Ecology* 22: 1–16, 1997.

[71] T. A. Witten and L. M. Sander. Diffusion-limited aggregation, a kinetic critical phenomenon. *Phys. Rev. Lett.* 47: 1400–1403, 1981.

[72] G. Wolf (ed.). *Encyclopaedia Cinematographica: Microbiology.* Institut für Wissenschaftlichen Film, Göttingen, 1967.

[73] E. Donchin, W. Ritter, and W. C. McCallum. Cognitive psychophysiology: the endogenous components of the ERP. In *Event-Related Potentials in Man.* (ed. E. Callaway, P. Tueting and S. H. Koslow). New York, Academic Press, 1978, pp. 349-412.

[74] G. Dumermuth and L. Molinari. Spectral analysis of EEG background activity. In *Methods of Analysis of Brain Electrical and Magnetic Signals, Handbook of Electroencephalography and Clinical Neurophysiology* (ed. Gevins, A.S. and Remond, A.) Revised series, Elsevier, Amsterdam, Vol. 1., 1987, pp. 85–130.

[75] E. Niedermeyer and F. Lopes da Silva, (ed.). *Electroencephalography.* Urban and Schwarzenberg, New York, 1993.

[76] A. S. Gevins and A. Remond (ed.). *Handbook of Electroencephalography and Clinical Neurophysiology*, Revised Series, Elsevier, Amsterdam, 1987.

[77] W. J. Freeman, Characterization of state transitions in spatially distributed, chaotic, nonlinear, dynamical systems in cerebral cortex. *Integrative Physiological and Behavioral Science* 29: 294-306, 1994.

[78] P. L. Nunez, *Neocortical Dynamics and Human EEG Rhythms*, Oxford University Press, Oxford, 1995.

[79] F. Lopes da Silva, Dynamics of EEGs as signals of neuronal populations: models and theoretical considerations. In *Electroencephalography.* (ed. Niedermeyer, E. and Lopes da Silva, F.). Urban and Schwarzenberg, New York, 1993.

[80] H. G. Schuster, *Deterministic Chaos.* Weingheim; Basel, 1988.

[81] R. W. Adey, Organization of brain tissue: is the brain a noisy processor? *J. Neuroscience* 3: 271–284, 1972.

[82] T. Elbert, W. J. Ray, Z. J. Kowalik, J. E. Skinner, K. E. Graf and N. Birbaumer. Chaos and physiology: deterministic chaos in excitable cell assemblies. *Physiological Reviews* 74: 1–47, 1994.

[83] W. S. Pritchard and D. W. Duke. Measuring 'chaos' in the brain: a tutorial review of EEG dimension estimation. *Brain and Cognition* 27: 53-397, 1995.

[84] A. Babloyantz. Chaotic dynamics in brain activity. In *Chaos in Brain Function* (ed. E. Basar). Springer, Berlin, 1990, pp. 42–48.

[85] P. Grassberger and I. Procaccia. Measuring the strangeness of strange attractors. *Physica D* 9: 189-208, 1983.

[86] C. J. Stam, T. C. A. M. van Woerkom, and R. W. M. Keunene. Non-linear analysis of the electroencephalogram in Creutzfeldt–Jakob disease. *Biological Cybernetics* 77: 247-256, 1997.

[87] A. Babloyantz and A. Destexhe. Low-dimensional chaos in an instance of epilepsy. *Proc. Natl. Acad. Sci.*, USA 83: 3513-3517, 1986.

[88] K. Lehnertz and C. E. Elger. Spatio-temporal dynamics of the primary epileptogenic area in temporal lobe epilepsy characterized by neuronal complexity loss. *Electroencephalography and Clinical Neurophysiology* 95: 108-117, 1995.

[89] J. E. Skinner, A. L. Goldberger, G. Mayer-Kress, and R. E. Ideker. Chaos in the heart: implications for clinical cardiology. *Biotechnology* 8: 1018-1024, 1990.

[90] J. E. Skinner, M. Molnár and C. Tomberg. The point correlation dimension: performance wih nonstationary surrogate data and noise. *Int. Physiol. and Behav. Sci.* 29: 217-237, 1994.

[91] J. Wackermann, K. Lehmann, I. Dvorak, and C. M. Michel. Global dimensional complexity of multichannel EEG indicates a change of human brain functional change after a single dose of a nootropic drug. *Electroencephalography and Clinical Neurophysiology* 86: 193-198, 1993.

[92] J. Röschke and J. B. Aldenhoff. A nonlinear approach to brain function: deterministic chaos and sleep EEG. *Sleep* 15: 95–101, 1992.

[93] J. Röschke and E. Basar. Correlation dimensions in various parts of cat and human brain in different states. In *Chaos in Brain Function* (ed. E. Basar). Springer–Verlag, Berlin, 1990, 92–109.

[94] M. Molnár and J. E. Skinner. Correlation dimension changes of the EEG during the wakefulness-sleep cycle. *Acta Biochimica et Biophysica Hungarica* 26: 121–125, 1991/92.

[95] J. Fell, J. Röschke, K. Mann, and C. Schäffner, Discrimination of sleep stages: a comparison between spectral and nonlinear EEG measures. *Electroenceph. Clin. Neurophysiol.* 98: 401–410, 1996.

[96] M. Mölle, L. Marshall, W. Lutzenberger, R. Pietrovsky, H. L. Fehm, and J. Born. Enhanced dynamic complexity in the human EEG during creative thinking. *Neurosci. Lett.* 208: 61–64, 1996.

[97] P. E. Rapp, T. R. Bashore, J. M. Martineire, A. M. Albano, I. D. Zimmermann, and A. I. Mees. Dynamics of brain electrical activity. *Brain Topography* 2: 99–118, 1989.

[98] M. Molnár, J. E. Skinner, V. Csépe, I. Winkler, and G. Karmos. Correlation dimension changes accompanying the occurrence of the mismatch-negativity and the P3 event-related potential component. *Electroenceph. Clin. Neurophysiol.* 95: 118–126, 1995.

[99] M. Molnár. The dimensional complexity of the P3 event-related potential: area-specific and task-dependent features. *Clinical Neurophysiology* 110: 31–38, 1999.

[100] R. A. M. Gregson, E. A. Campbell, and R. G. Gates. Cognitive load as the determinant of the dimensionality of the electroencephalogram: a replication study. *Biological Psychology* 35: 165–178, 1992.

[101] C. J. Stam, T. C. A. M. van Woerkom, and W. S. Pritchard. Use of nonlinear EEG measures to characterize EEG changes during mental activity. *Electroenceph. Clin. Neurophysiol.* 99: 214–224, 1996.

[102] J. E. Skinner, M. Molnár, T. Vybiral, and M. Mitra. Application of chaos theory to biology and medicine. *Int. Physiol. and Behav. Sci.* 27: 39–53, 1992.

[103] W. J. Freeman. The physiology of perception. *Sci. Am.* 264: 78–92, 1991.

[104] W. Lutzenberger, T. Elbert, W. J. Ray, and N. Birbaumer. The scalp distribution of the fractal dimension of the EEG and its variation with mental tasks. *Brain Topography* 5: 27-34, 1993.

[105] G. Sammer. Working-memory load and dimensional complexity of the EEG. *Int. J. Psychophysiology* 24: 173–182, 1996.

[106] E. Donchin and M. G. H. Coles. Is the P300 component a manifestation of context updating? *Behavioral Brain Sciences* 11: 357–374, 1988.

[107] R. Verleger. Event-related potentials and cognition: a critique of the context updating hypothesis and an alternative interpretation of the P3. *Behavioral Brain Sciences* 11: 343-427, 1988.

[108] M. Molnár. On the origin of the P3 event-related potential component. *Int. J. Psychophysiology* 17: 129–144, 1994.

[109] J. S. Buchwald. Comparisons of sensory and cognitive brain potentials in the human and in an animal model. In *Spinger Series in Brain Dynamics*. (ed. Basar, E. and Bullock, T.). No. 2, Springer–Verlag, Berlin-Heidelberg, 1989, pp. 242–257.

[110] J. S. Buchwald. Animal models of cognitive event-related potentials. In *Event-Related Potentials: Basic Issues and Applications*. (ed. J. W. Rohrbaugh, R. Parasuraman and R. Johnson). Oxford University Press, 1990, pp. 57–75.

[111] M. Molnár, G. Karmos, V. Csépe, and I. Winkler. Intracortical auditory evoked potentials during classical aversive conditioning in cats. *Biological Psychology* 26: 339–350, 1988.

[112] J. P. Pijn, J. V. Neerven, A. Noest, and F. Lopes da Silva. Chaos or noise in EEG signals: dependence on state and brain site. *Electroencephalography and Clinical Neurophysiology* 79: 371–381, 1991.

[113] K. Lehnertz and C. E. Elger. Can epileptic seizures be predicted? Evidence from nonlinear time series analysis of brain electrical activity. *Phys. Rev. Lett.* 80: 5019–5022, 1998.

[114] M. Molnár, Gy. Gács, G. Ujvári, J. E. Skinner, and G. Karmos. Dimensional complexity of the EEG in subcortical stroke – a case study, *Int. J. Psychophysiology* 25: 193–199, 1997.

5 Microscopic mechanisms of biological motion

I. Derényi and T. Vicsek

The origin of biological motion can be traced back to the function of tiny *motor proteins* operating at the *molecular scale* [1,2]. During the last few decades numerous different families of these molecular motors have been discovered, and the newest experimental techniques developed in recent years have allowed us to study *in vitro* the operation of *individual* motor proteins (for a review see [3]). Such studies have revealed that the motors move in discrete, unidirectional steps with step sizes in the range of a few nanometres and exert piconewton forces. The motors use the *chemical energy* stored in adenosine triphosphate (ATP) molecules or in a proton gradient as fuel and convert it into *mechanical work*.

5.1 Characterization of motor proteins

5.1.1 Cytoskeleton

Eukaryotic cells contain a dynamic protein network, the *cytoskeleton* (Fig. 5.1), which maintains the shape of the cell, anchors organelles, moves the cell, and serves as highways for intracellular transport. It consists of three kinds of filamentous structures: the *microtubules*, the *actin filaments* (also known as microfilaments) and the *intermediate filaments*. Along these polymeric filaments molecular motor proteins can move, consuming the chemical energy stored in ATP molecules. Three different motor protein families have now been identified: *kinesins*, *dyneins* which move along the microtubules, and *myosins* which move along the actin filaments. There are no (known) motor proteins associated with the intermediate filaments.

Microtubules are straight, hollow tubes (24 nm in diameter), the wall of which is made up of tubulin heterodimers arranged in 13 longitudinal rows called protofilaments (Fig. 5.2). A tubulin heterodimer is 8 nm long and consists of two globular proteins about 4 nm in diameter: α- and β-tubulin. The dimers bind head-to-tail giving the polarity to the protofilaments. The microtubule has a helical surface lattice with two possible configurations: the adjacent protofilaments can be shifted in the longitudinal direction by about either 5 nm (A-type lattice) or 1 nm (B-type lattice).

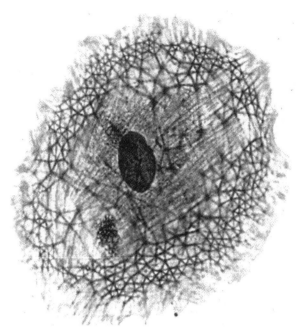

Fig. 5.1 The cytoskeleton. A cell has been fixed and stained with a general stain for proteins. A variety of filamentous structures that extend through the cell can be seen (after [2]).

Electron micrograph measurements of microtubules decorated with kinesin head fragments [4–7] indicate the predominance of the B-type lattice and, in addition, show that kinesin heads can bind only to the β-tubulin (meanwhile weakly interacting with the α-tubulin as well).

In most cells, the 'minus' (slow growing) ends of the microtubules are stabilized and embedded in a structure near the nucleus, called the centrosome. In addition to the structural role of the microtubules, they serve as tracks for the microtubule associated motor proteins that carry organelles (such as mitochondria; see Fig. 5.3 from [8]), vesicles (e.g., from the Golgi apparatus to their destination) or large molecules (chromosomes during cell division) inside the cytoplasm. Kinesins usually walk towards the 'plus' (fast growing) end of the microtubules, while dyneins walk towards the minus end. Nonclaret disjunctional (ncd) proteins, which are kinesin-related microtubule motor proteins, are also minus end motors.

Microtubules play the major role in the beating of eukaryotic cilia and flagella. The basis of these external cellular structures is formed by a ring of nine doublet microtubules, with two singlet microtubules at the centre. A doublet microtubule consists of 23 protofilaments in a figure-of-eight arrangement. The peripheral doublets are connected by dyneins. The motion of these dyneins results in a relative sliding of adjacent doublets, inducing the cilia and flagella to bend.

Actin filaments are assembled from the globular protein actin and are the thinnest (7 nm in diameter) of the cytoskeleton fibres. Individual actin monomers polymerize

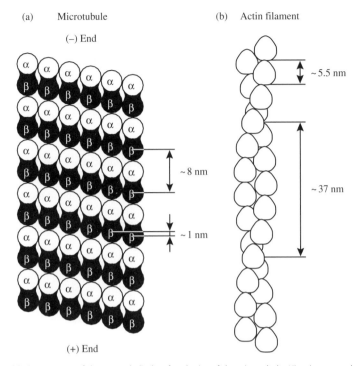

(a)   Microtubule

(−) End

(+) End

(b)   Actin filament

~5.5 nm

~8 nm

~37 nm

~1 nm

Fig. 5.2 (a) The structure of the B-type helical surface lattice of the microtubule. Kinesin moves along the protofilaments (vertical rows) and its heads bind mainly to the β monomers (after [69]). (b) The double-helical structure of the actin filament.

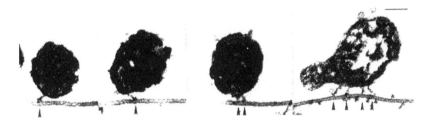

Fig. 5.3 The picture taken by electron microscopy shows some mitochondria (large objects) as they are carried by several motor proteins (indicated by arrow heads) along a microtubule (after [8]). Reprinted by permission from Nature, copyright 1990 Macmillan Magazines Ltd.

into a chain and two chains twist together to form a helix with 13.5 molecules (74 nm) per turn (Fig. 5.2). Many actin filaments are attached to proteins within the plasma membrane. By this attachment, the filaments can help maintain the shape of cells. Where the actin interacts with its motor protein, the myosin, the generated force can change the shape of the cell, allowing it to put out pseudopodia (for amoeboid movement), to invaginate particles (during phagocytosis), to cleave the cell (during cell division), or to cause *muscle contraction* (see §5.1.2 for details). This system also plays some role in intracellular transport processes.

Intermediate filaments are intermediate in size (8–12 nm in diameter) between microtubules and actin filaments. They are made of different kinds of proteins in different types of cells. While microtubules and actin filaments can be polymerized and depolymerized rapidly, intermediate filaments tend to be more stable and help maintain the shape of the cell over long periods of time. The nucleus is held in place in the cytoplasm by a particular form of intermediate filament.

Kinesin

Kinesin was discovered in the mid-1980s [9]. Further studies have led to the identification of a large number of proteins that are related in structure to kinesin and constitute the kinesin superfamily of motor proteins. Native kinesin is a dimeric molecule with a tail and two globular ($\sim$9×3×3 nm) head domains [10, 11] (Fig. 5.4(a)). The heads are highly conserved, and each contains an ATP and a tubulin binding site. The more or less extended tail interacts with the cargo.

Recent experimental studies of individual kinesin molecules in *in vitro* motility assays (see §5.1.4) have revealed the following properties of kinesin movement:

- Kinesin moves unidirectionally parallel to the protofilaments towards the plus end of the microtubule [9, 12].

- Under an increasing load the speed of the kinesin decreases almost linearly [13, 14] (see Fig. 5.15).

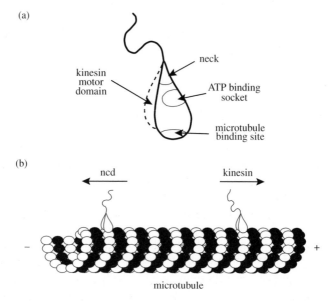

Fig. 5.4 (a) Schematic drawing of the two-headed kinesin molecule. (b) Kinesin and ncd are structurally very similar proteins and bind on the microtubule with similar orientations; yet, they move in opposite directions.

- Under its stall force (about 5 pN) kinesin still consumes ATP at an elevated rate [14].

- In the absence of ATP (in rigour state) kinesin binds to the microtubule very strongly: it supports forces in excess of 10 pN [14].

- The step size ($\sim$ 8 nm) is identical with the periodicity of the protofilaments [15, 16].

- Kinesin hydrolyses one ATP per each 8 nm step [17].

- Occasionally kinesin takes backward steps [14, 16].

- The displacement variance at saturating ATP and at low load increases linearly with time, but at (only somewhat more than) half of the rate of a single Poisson stepping process with 8 nm step size, implying that one step consists of two sequential subprocesses with comparable limiting rates [18, 19].

Ncd, which is also a member of the kinesin superfamily, moves in the opposite direction, towards the minus end of the microtubule (Fig. 5.4(b)). This is a surprising phenomenon, because the motor domains of kinesin and ncd are structurally very similar and bind on the microtubule with similar orientations [11]. This mystery has been deepened by a recent elegant experiment [20], in which a chimera was formed by attaching the motor domain of ncd to the neck region of kinesin. Surprisingly, the resulting motor catalysed plus-end-directed motion characteristic of kinesin from which the neck (and not the motor) region was taken. Attaching the motor domain of kinesin to the neck region of ncd [21] led to similar results: the directionality of the chimera was determined by the neck region. Moreover, changing a few amino acids in the neck can reverse the directionality of the motor [21, 22]. Structural studies also show that it is, in the neck region where kinesin and ncd differ the most [22].

5.1.2 Muscle contraction

Muscle is an organ specializing in force generation and movement at the macroscopic scale. There are many types of muscles. Vertebrate muscles fall into three categories: skeletal muscle (or striated muscle; see Fig. 5.5), responsible for locomotion, flight, etc.; cardiac muscle, which has a vital role and is able to function for a century or more without ever taking a break; and smooth muscle (or involuntary muscle) which lines the walls of the arteries to control blood pressure, or controls the digestion of food by causing movement of the intestine. Some insect flight muscles are adapted for speed and can contract 1000 times a second, while molluscan adductor muscles are built for strength and can lift a 10 kg weight per square centimetre cross-section. In spite of this large diversity the basic mechanism of force generation and contraction is common in all types of muscles: *myosin cross-bridges* extending from the thick myosin filaments attach to the binding sites of the thin actin filaments and exert force on them, inducing a relative sliding of the actin and myosin filaments. This process is fuelled by the hydrolysis of ATP molecules.

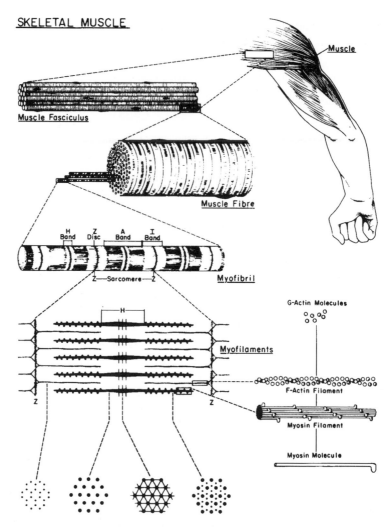

Fig. 5.5 The structure of the vertebrate skeletal muscle (after [23]).

Myosin cross-bridge

Myosins, similarly to kinesins, are also dimeric molecules with a long tail and two globular heads (cross-bridges), but probably only one of the two heads can interact with the actin filament at a time. Each head contains an ATP and an actin binding site. The tails of many myosins attach to each other, forming the thick myosin filament. The mechanical force and motion is generated when a head which is bound to the actin undergoes a conformational change. Though individual heads usually spend only a small fraction of their time strongly bound to actin, the large number of cross-bridges between the actin and myosin filaments prevents them from diffusing away.

The widely accepted cross-bridge scheme for the states of the actomyosin ATPase, which was originally proposed by Lymn and Taylor [24], is

$$A \cdot M^* \cdot ADP \quad \xrightleftharpoons{\text{power-stroke}} \quad A \cdot M \cdot ADP \quad \xrightleftharpoons[k_{ADP+}]{k_{ADP-}} \quad A \cdot M$$

$$P_i \diagdown \Big\Updownarrow {}^{k_{P+}}_{k_{P-}} \qquad\qquad\qquad\qquad \Big\Updownarrow {}^{ATP}_{k_{ATP+}}$$

$$(A) \cdot M \cdot ADP \cdot P_i \quad \xrightleftharpoons{\substack{ATP- \\ \text{hydrolysis}}} \quad (A) \cdot M \cdot ATP$$

(For review see [25,26].) Initially, the myosin cross-bridge is in a weakly bound, non-force-producing $(A) \cdot M \cdot ADP \cdot P_i$ state. A and M denote actin and myosin, respectively, and (A) indicates that myosin is weakly bound to actin. Releasing the phosphate P_i the cross-bridge gets into a strongly bound, force-producing $A \cdot M^* \cdot ADP$ state. Then the cross-bridge executes a *power-stroke* (see Fig. 5.19(a)) producing mechanical force on actin, a conformational change in myosin, and a relative sliding between adjacent actin and myosin filaments. This new $A \cdot M \cdot ADP$ state cannot readily bind P_i any more, in contrast to the preceding one. At the end of the power-stroke, when continued attachment would resist useful work, ADP is released with transition to the $A \cdot M$ state, which is still strongly bound. ADP may then rebind, resulting in continued counter-productive attachment, or ATP may bind, resulting in rapid cross-bridge detachment. The hydrolysis step occurs in this weakly bound $(A) \cdot M \cdot ATP$ state resulting in the $(A) \cdot M \cdot ADP \cdot P_i$ cross-bridge state, which is free to participate in a new working cycle.

Recent *in vitro* experimental studies of individual myosin molecules have revealed that a cross-bridge can generate a force of up to 3–4 pN, and the power-stroke working distance is about 11 nm [27]. This distance is also consistent with the structural studies of myosin [28–31]. The conformations of the cross-bridge states was determined by Electron Paramagnetic Resonance (EPR) spectroscopy [32]. These data together with fluorescence polarization measurements [33] indicate that the 11 nm-long power-stroke is the result of the tilt and twist of the elongated neck (light-chain) domain of the myosin's head.

5.1.3 Rotary motors

The motor proteins discussed above (kinesin, ncd, dynein, myosin) can be called *linear motors*, since they move along straight, one-dimensional structures. In contrast, *rotary motors* do not advance large distances, but they rotate their shaft (or rotate around their shaft).

One example is the *bacterial flagellar motor* which is anchored in the cell membrane and rotates its flagellar filament that propels the cell. The motor can rotate in both directions and is fuelled by the electrochemical gradient between the two sides of the membrane. This gradient is typically a proton (H^+) gradient, but it can also be accomplished using sodium ions (Na^+), particularly by bacteria living in highly alkaline environments where there are few protons, or in marine environments where there is an abundance of sodium ions.

Other motors (typically DNA motors) move unidirectionally along their polymeric track, while they also rotate around it. Thus, their overall motion is helical rather than linear.

ATP synthase

Perhaps the most peculiar rotary motor system is the enzyme *ATP synthase* which is present in the membranes of bacteria, chloroplasts and mitochondria, and has been remarkably conserved through evolution. The function of the ATP synthase is to couple the synthesis of ATP with a proton flux across the membrane down the proton gradient. The flux goes from the protochemically positive P-side (high proton electrochemical potential) to the protochemically negative N-side. In bacteria the P-side is the outside and the N-side is the cytoplasm; in mitochondria the P-side is the intermembrane space and the N-side is the mitochondrial matrix; in chloroplasts the P-side is the lumen and the N-side is the stroma. The operation of the ATP synthase is fully reversible; it can generate proton flux up the proton gradient by ATP hydrolysis. In some bacteria this is the main function of the enzyme.

The ATP synthase is a mushroom-shaped enzyme (see Fig. 5.6), consisting of two parts: the F_0 portion (stalk), which is the proton channel that spans the membrane, and the F_1 portion (button), which is a cytoplasmic domain containing the active catalytic sites. Both parts are *rotary molecular motors* working against each other via their common shaft. The F_0 part is driven by the proton gradient and the F_1 part by the ATP hydrolysis, and the 'stronger' motor forces the 'weaker' one to rotate in the reverse direction. This is how ATP synthase is able to synthesize ATP molecules from ADP and P_i consuming the energy stored in the proton gradient or vice versa [34–37].

The F_0 portion consists of 3 subunits in a stoichiometry of 1a : 2b : 9–12c. The twin α-helix-shaped c-subunits form a cylinder, which is believed to be the 'rotor', while the a-subunit consists of 5 transmembrane helices, and constitutes the 'stator'. Each of the c-subunits has a conserved acidic residue half-way across the membrane. The idea for the operation of the F_0 motor suggested by Wolfgang Junge [37] is that the a-subunit provides a port for entry of the protons from the P-side and a port for exit to the N-side. When a proton enters through the P-size port, it neutralizes the acidic residue of the c-subunit. Only in this neutral form can the c-subunit complex rotate away from association with the a-subunit. Rotation brings a neutral c-subunit to the exit port, allowing it to lose the proton and associate with the a-subunit. Successive protonations allow the c-subunit complex to rotate by $1/N_c \times 360°$ for each proton, where $N_c = 9 - 12$ is the stoichiometry of the c-subunit per ATP synthase.

The F_1 portion (which is also known as F_1 ATPase) contains 5 subunits, in a stoichiometry of $3\alpha : 3\beta : 1\gamma : 1\delta : 1\varepsilon$. The α- and β-subunits in alternating order form a

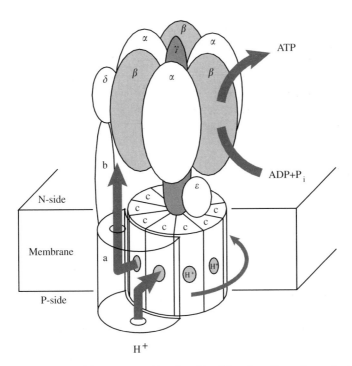

Fig. 5.6 Schematic picture of the structure and function of the ATP synthase. The proton gradient across the membrane induces the rotation of the F_0 component (lower part) of the enzyme, which then forces the F_1 component to synthesize ATP from ADP and P_i.

hexagonal ring (stator), surrounding the γ-subunit (rotor or shaft). The 3 active catalytic ATP binding sites are in the β-subunits, and the additional nucleotide binding sites in the α-subunits are regulatory. The F_1 portion alone catalyses ATP hydrolysis, but not ATP-synthesis. It operates through a mechanism in which the three active sites undergo a change in binding affinity for the reactants of the ATPase reaction, ATP, ADP and P_i. The change in affinity accompanies a change in the position of the γ-subunit relative to the α-β-ring, which involves a rotation of the one relative to the other [38, 39], and a 120° phase-shift in the catalytic cycle of the 3 active sites relative to one another. This so-called *binding change mechanism* (Fig. 5.24), which was first proposed by Paul Boyer [34], is confirmed by two direct pieces of evidence: (i) John Walker's group solved the crystal structure of the F_1 ATPase [40], and found that the 3 active sites are indeed in 3 different states of the catalytic cycle; and (ii) the rotational motion was captured in dramatic videos from the laboratory of Masasuke Yoshida [39] by tethering the α-β-ring to a glass surface and detecting the rotation of a fluorescently labelled actin filament that was attached to the γ-subunit.

The exact role of the b-, δ- and ε-subunits is unknown; they are probably structural elements: the b- and δ-subunits may link the stators of the F_0 and F_1 portions, preventing the α-β-ring from moving, while the ε-subunit may help connecting the rotors.

5.1.4 Motility assay

In vitro motility assays (e.g., see [41, 42]) have truly revolutionized the study of the operation of linear motor proteins. Purified motors can be examined as they move along a single complementary filament (actin or microtubule) in the presence of controlled amounts of substances (such as ATP) and in the complete absence of other components of the cell. In such systems both the biochemical (such as ATP consumption) and mechanical (force, velocity, step size) properties of motor proteins can be measured quantitatively. In addition, kinesin and cytoplasmic dynein were identified also in motility assays. On the other hand, the cellular role of numerous motors is still undetermined, for the most part.

In a conventional motility assay a quartz or glass surface is coated with motor proteins, and then a solution of the complementary filaments is added. The motors grab the filaments and move them unidirectionally across the surface (Fig. 5.7). If the concentration of the motors is low enough, the motion of single molecules can be discerned. The motion of fluorescently labelled filaments can be observed directly by using fluorescent microscopy. Two possible ways of controlling the load force against which the motors work are: (i) attaching a microneedle to the filament; or (ii) applying a viscous drag force to the moving filament, increasing the viscosity of the medium for the filament by about 100-fold.

A great advance in the experiments was the use of *optical traps* (sometimes called *optical tweezers*) to facilitate the examination of *individual* motor proteins.

Optical trap

The optical trap uses a highly focused laser light to grab and manipulate microscopic dielectric objects. Its unique trapping capability come from the three-dimensional gradient in light intensity found near the focus. When a moderately powerful laser is focused to a diffraction-limited spot in the specimen plane of a microscope, a steep

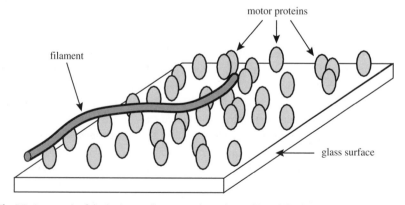

Fig. 5.7 An example of the *in vitro* motility assays where a lawn of immobilised motor proteins moves a complementary filament.

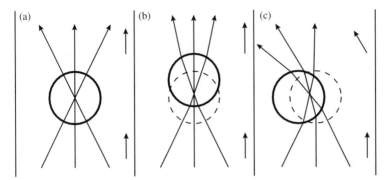

Fig. 5.8 Illustration of the trapping capability of an optical trap by tracing three typical rays of a highly focused laser light. (a) The equilibrium position of the bead. (b) If the bead is displaced out of its equilibrium position in the longitudinal direction, the bead transfers momentum to the light because of the refraction, resulting in a force that draws the bead back to its equilibrium position. The total momentum of the light is indicated by small arrows on the right-hand side of the figures. (c) If the bead is displaced out of its equilibrium position in the transversal direction, the force arising from the momentum transfer also draws back the bead.

gradient in light intensity is produced in the focal region. Small dielectric objects, such as latex or silica micro-spheres, or biological material, including cells, chromosomes, and organelles, experience a form of radiation pressure, called the gradient force, that tends to draw them towards the centre of that region, where their dielectric energy is minimal. The dielectric energy density in the object is

$$\rho(x) = -\tfrac{1}{2}\Delta\alpha E(x)^2, \tag{5.1}$$

where $E(x)$ is the electric field of the light, and $\Delta\alpha$ is the polarizability of the object relative to the suspending solution. Consequently, this energy density is minimal in the centre of the focal region, where the light has the maximum intensity.

For a spherical dielectric object (bead) the gradient force can also be understood easily from the law of conservation of momentum. Figure 5.8 illustrates that if the bead is displaced out of its equilibrium position, the total momentum of the light changes during the refraction, and this momentum transfer between the bead and the light results in an effective force that draws the bead back to its equilibrium position.

Another form of radiation pressure, usually called the scattering force, arises from reflection or absorption of light, and it tends to push objects down along the beam in the direction of the propagation of the light. Stable, three-dimensional trapping takes place when the effect of the gradient force is sufficiently large to overcome the scattering force. In practice, such a condition can be achieved by using microscope objectives with the highest possible numerical aperture, which generate the steepest gradients.

The laser beam can serve not only to confine small dielectric objects, but also to measure their position. In addition, because the optical trap behaves like a spring, i.e., the trap force grows in proportion to the displacement of the object from the centre, the force acting on the object can also be determined.

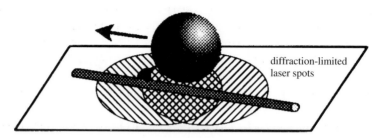

Fig. 5.9 A kinesin molecule walks along an immobilized microtubule and carries a silica bead which is grabbed by an optical trap. The displacement of the bead is sensed optically (after [15]). Reprinted by permission from Nature, copyright 1993 Macmillan Magazines Ltd.

Fig. 5.10 An immobilized myosin molecule tries to pull the actin filament, the ends of which are attached to polystyrene beads and held in place by optical traps (after [27]). Reprinted by permission from Nature, copyright 1994 Macmillan Magazines Ltd.

Svoboda *et al.* [15] used first this techniques in motility assay. They added kinesin to 0.6 µm silica beads at a concentration so low that, on average, less than one molecule of kinesin clung to each bead. Then they coated a substratum with microtubules and placed a kinesin-carrying bead on a microtubule, using the optical tweezers, and measured the displacement of the bead that was pulled by the kinesin molecule (Fig. 5.9). The results of this experiment have been discussed in detail in §5.1.1.

To measure the mechanical properties of single myosin molecules, which remains anchored to actin for only a small fraction of the stepping cycle, Finer *et al.* [27] used optical traps to capture the filaments rather than the motor molecules (Fig. 5.10). Their main results have been summarized in §5.1.2.

5.2 Fluctuation-driven transport

The wide range of new experimental results on motor proteins and the desire to understand the basic mechanism of their operation have led to the development of a new statistical physical principle, called *fluctuation-driven transport*. It is based on the discovery that nonequilibrium fluctuations, whether generated externally or by a chemical reaction far from equilibrium, can drive directed motion along a periodic and spatially asymmetric structure (ratchet), without thermal gradients or net macroscopic forces, simply by biasing Brownian motion. Systems operating on this principle are often referred to as *thermal ratchets* or *Brownian ratchets* (for a review see [43–47]). Since living cells maintain their ATP concentration far from equilibrium, the ATP hydrolysis can be considered as the source of nonequilibrium fluctuations for most motor proteins, and the polymeric filaments along which these motors move are the required periodic and asymmetric (polar) structures.

On symmetry grounds breaking the temporal and spatial symmetry simultaneously is enough for the existence of a velocity vector. In a Brownian ratchet system temporal symmetry is automatically broken because of the dissipative (viscous) environment, and the spatial symmetry is broken by the asymmetry (polarity) of the structure. A simple choice for this structure – inspired by Feynman's ratchet – is a sawtooth-shaped periodic potential (Fig. 5.11(a)). In his *Lectures* [48], Feynman used the 'ratchet and pawl' engine to illustrate some implications of the second law of thermodynamics and, in particular, that useful work cannot be extracted from equilibrium fluctuations (thermal noise). This justifies that, indeed, additional nonequilibrium fluctuations are necessary for a Brownian motion to be rectified, resulting in net motion (and work if some external load is attached to the particle). The two basic types of nonequilibrium fluctuations are the fluctuating potential (or 'flashing ratchet') [49–52], where the potential is time dependent and fluctuates between two or more different states, and the fluctuating force (or 'rocking ratchet') [50, 52, 53], where the potential is static but a fluctuating external force with zero time average is applied.

The Langevin equation (2.23) for the motion of a particle in such systems can be written as

$$\gamma \dot{x} = -\partial_x V(x, t) + F(t) + F_{\text{load}} + \gamma \sqrt{2D} \xi(t), \tag{5.2}$$

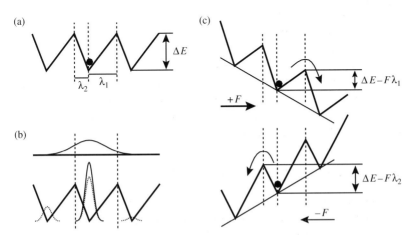

Fig. 5.11 The motion of a Brownian particle (a) in a sawtooth-shaped potential due to the effect of (b) a fluctuating potential and (c) a fluctuating external force (after [47]). Reprinted with permission from Elsevier Science, copyright 1998.

where $V(x, t)$ denotes the periodic ratchet potential (which can fluctuate in time), $F(t)$ is the fluctuating external force (if there is any), and F_{load} is the load force if the particle has to move against some external load.

5.2.1 Basic ratchet models

Figure 5.11(b) is an example for the flashing ratchets and illustrates the time evolution of the probability density of the position of a particle when a sawtooth potential is switched on and off repeatedly at an appropriate switching rate. When the sawtooth potential is switched on and the temperature is small enough compared to the height of the barriers, the particle stays near the bottom of a well. After switching off the sawtooth potential the particle starts to diffuse on the flat potential. Switching back to the sawtooth potential, the particle has a larger probability of being captured by the neighbouring well on the left than by the neighbouring well on the right because of the asymmetry of the potential. Thus, a net motion to the left can be observed.

Figure 5.11(c) is an example for the rocking ratchets, where an external force alternates between $+F$ and $-F$ with an appropriate characteristic frequency with zero time average. When the force points to the right ($F(t) = +F$), the energy barrier for jumping to the right becomes shorter by F-times the longer distance λ_1. When the same force points to the left ($F(t) = -F$), the energy barrier for jumping to the left becomes shorter but only by F-times the shorter distance λ_2. Thus, a net motion to the right is expected.

The frequency response for the two types of fluctuations is very different [50] (Fig. 5.12). The fluctuating force causes a net flow at low frequencies. With increasing

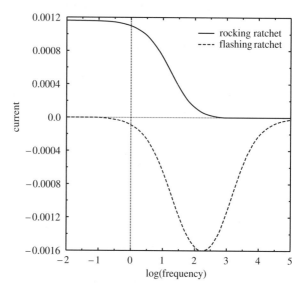

Fig. 5.12 An example of the net current in a rocking and a flashing ratchet as a function of the frequency of the fluctuations.

frequency the flow decreases slowly but, exceeding the inverse of the intrawell relaxation time, the flow approaches zero rapidly. On the other hand, for fluctuating potential a maximum flow occurs in an intermediate frequency range and the flow vanishes at both high and low frequencies.

5.2.2 Brief overview of the models

The transport induced by fluctuating forces has recently been studied for various statistics of the fluctuation [54–58]. In several cases a surprising phenomenon, the reversal of the direction of the motion, has been observed for changing some parameters of the statistics. The spatial symmetry of the system can be broken not only by the asymmetry of the potential, but an asymmetrically fluctuating force (with zero mean) can also result in net flow even if the potential is spatially symmetric [59–61].

In the case of the fluctuating potential the transition rate constants between the different potentials (states) may depend on the position of the particle, and the spatial asymmetry can be embedded not only in the shape of the potential, but also in the transition rates. The most efficient transport with fluctuating potentials can be achieved when the extrema of the two (or more) potentials are shifted relative to each other, and a continuous downslope motion of the particle can be attained [62] (Fig. 5.13). If the potential fluctuates between two states that are the inverses of one another, the direction and velocity of the motion are very sensitive to the shape of the potential, the time characteristics of the fluctuations and some externally applied small forces [63,64].

Nonuniform periodic temperature distribution alone can also induce net flow (the 'blow torch effect' of Landauer [65–67]), because the system is out of equilibrium.

5.2.3 Illustration of the second law of thermodynamics

At the beginning of this chapter it has been argued that thermal noise alone can never produce rectified motion in a ratchet potential because of the second law of thermodynamics. This can also be proven easily for an overdamped Brownian particle

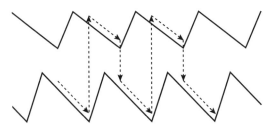

Fig. 5.13 An example how a continuous downslope motion can be attained if the transition between the two potentials is allowed only in the vicinity of the minima, and the extrema of the potentials are shifted relative to one another.

moving in an arbitrary one-dimensional periodic potential $V(x)$ with period λ. In this case the motion of the particle is described by the Fokker–Planck equation (2.24) with probability current eqn (2.26). Since the potential $V(x)$ is periodic, it is enough to consider the motion in one period ($0 \leq x < \lambda$) with periodic boundary conditions. In the steady state, when the probability density $P(x, t)$ and the current $J(x, t)$ have already reached their stationary values $P_{st}(x)$ and $J_{st}(x)$, the Fokker–Planck equation leads to $\partial_x J_{st}(x) = 0$, i.e., the stationary current is constant in space: $J_{st}(x) = J_{st}$. After rearranging and integrating eqn (2.26) from 0 to λ in the steady state we get

$$J_{st} \frac{\gamma}{k_B T} \int_0^\lambda e^{V(x)/k_B T} \, dx = - \left[e^{V(x)/k_B T} P_{st}(x) \right]_{x=0}^\lambda . \tag{5.3}$$

Since the period of the system is λ, the right-hand side of this equation is zero. Because the integral on the left-hand side is always positive, independently of the potential $V(x)$, the stationary current must also be zero:

$$J_{st} = 0, \tag{5.4}$$

which means that indeed, no net motion can be gained from thermal fluctuations. Note that the stationary (or equilibrium) solution of the probability density is the Boltzmann distribution:

$$P_{st}(x) = \frac{e^{-V(x)/k_B T}}{\int_0^\lambda e^{-V(x)/k_B T} \, dx}, \tag{5.5}$$

which follows from eqn (2.26) with zero current.

5.3 Realistic models

Although the fluctuation-driven transport sheds light on the basic mechanism of motion generation and energy conversion by molecular motors, specific models are needed to describe real biological systems.

5.3.1 Kinesin

There are two basic mechanisms for the stepping of the two-headed kinesin molecule [68]. The first one is the so-called 'Two-Step' model (see Fig. 5.14) in which both heads of the kinesin advance 8 nm in rapid succession for the hydrolysis of only one ATP by one of the heads. The second mechanism is the 'Long-Stride' model (see Fig. 5.17) in which the back head detaches, advances 16 nm passing by the bound front head, and reattaches to the nearest binding site in front of the bound head. Then the heads change their roles and a new step may take place. With both mechanisms the centre of the kinesin advances 8 nm for each step.

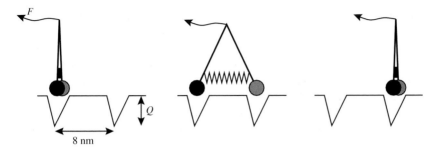

Fig. 5.14 Schematic picture of the ratchet potential and the two substeps of the kinesin molecule in the Two-Step model (time passes from left to right) (after [69]).

Two-Step model

Several versions of the Two-Step model can be distinguished with respect to the relative positions of the heads of the kinesin. If both heads track the same protofilament, they are not able to pass by each other, one of the heads always trails the other one, and their distance alternates between 8 and 16 nm during each step. If the heads track adjacent protofilaments, their relative position (looking from the side) alternates most probably between ≈ 0 and ≈ 8 nm, and the front head in a step may become the back head in another step.

The ratchet model [69] we present here describes the whole family of the Two-Step models. We suppose that each of the two Brownian heads moves along its own one-dimensional periodic potential with period $L = 8$ nm in an overdamped environment. These potentials represent the interaction between the heads and the protofilaments. Each period contains a deep potential well corresponding to the binding site of the β-tubulin and the potentials everywhere else are flat. Each well has an asymmetric 'V' shape (see Fig. 5.14) reflecting the polarity of the protofilaments: the slope in the backward direction (towards the minus end of the microtubule) is steeper and 0.5–1 nm long, while the other slope (towards the plus end) is 1.5–2 nm long. These ranges are of the order of the Debye length. The potentials for the two heads can be shifted relatively to each other by an arbitrary distance, depending on the version of the Two-Step model.

In this specific model the heads are connected at a hinge, and a spring acts between them [70] (Fig. 5.14). At the beginning of the mechanochemical cycle both heads are sitting in their wells waiting for an ATP molecule, and the spring is unstrained. Any configuration of the model is mathematically equivalent to the one in which the shift between the two potentials is zero, the heads are at the same position and, therefore, the rest length of the spring is zero. (Further on we will consider only this version of the model.) After one of the heads binds an ATP, its hydrolysis causes a conformational change in this head; more precisely, it induces the head to take a 8 nm forward step. In the language of the model this means that the rest length of the spring changes from zero to 8 nm right after the hydrolysis. Then, as the first rate-limiting subprocess, the strained spring tries to stretch, pushing one of the heads to the neighbouring well. Reaching its new 8 nm rest length another conformational change occurs in the head as a consequence of the ADP release: the rest length of the spring changes back to zero. Then, as the second rate-limiting subprocess, the spring tries to contract, pulling one of the heads to the position where the other one is located. Completing the contraction, a new cycle can begin. Due to the asymmetry of the potentials an expansion and contraction in most instances results in an 8 nm forward step. Hirose *et al.* [7] provided evidence that the kinesin·ADP complex has indeed a different conformation near the junction of the heads. The model is also consistent with the scheme of Gilbert *et al.* [71] for the pathway of the kinesin ATPase, and naturally explains the low displacement variance of the kinesin's motion [18, 19]. An important cooperative feature of this model is that only one ATP hydrolysis occurs during each cycle.

We can take the load force into account in a natural way. Since in the experiments of Block's group [14, 15] a large (∼0.5 μm in diameter) and therefore slow (compared to the kinesin heads) silica bead was linked to the hinge of the kinesin molecule by a relatively weak elastic tether, we can apply a constant force F to the hinge. We arrive at the same conclusion by considering the experiments of Hunt *et al.* [13], where a viscous load acted on the moving microtubule while the long tail of the kinesin was fixed.

The motion of the heads can be described by two coupled Langevin equations

$$\gamma \dot{x}_1 = -\partial_x V(x_1) - F_1^{\text{load}} - K\left[x_1 - x_2 - l(t)\right] + \gamma \sqrt{2D}\xi_1(t),$$
$$\gamma \dot{x}_2 = -\partial_x V(x_2) - F_2^{\text{load}} + K\left[x_1 - x_2 - l(t)\right] + \gamma \sqrt{2D}\xi_2(t),$$

where x_1 and x_2 denote the positions of the front and back heads respectively, $V(x)$ is the periodic ratchet potential, K is the stiffness of the spring, $F_1^{\text{load}} = F_2^{\text{load}} = F/2$ are the load forces acting on the heads, and the two thermal noise terms $\xi_1(t), \xi_2(t)$ are uncorrelated. The load force F can be distributed between F_1^{load} and F_2^{load} in different ways, but due to the robustness of the model they lead practically to the same outcome. $l(t)$ represents the rest length of the spring, which is not an explicit function of time, but alternates stochastically between 0 and 8 nm as ATP is hydrolysed or ADP is released.

In order to obtain results that we can compare with the experiments, first the values of the input parameters have to be specified. The friction coefficient γ for a single head can be calculated from the Stokes law, yielding $\gamma \approx 6 \times 10^{-11}$ kg/s. The free energy that can be gained from the ATP hydrolysis is about $25k_B T$ or 100×10^{-21} J. During a stepping cycle the two conformational changes consume the free energy of the ATP, as the rest length of the spring changes by 8 nm in both cases. This means that $25k_B T/2 \approx K (8 \text{ nm})^2/2$, from which we get $K \approx 1.5$ pN/nm.

A lower limit for the depth Q of the potential wells can be determined from the fact that the kinesin in rigour state supports forces in excess of 10 pN. If we try to pull out the two-headed kinesin molecule (with drag coefficient 2γ) from a $2Q$ deep potential well with a force F, a lower limit for the escape rate is $(F^2/k_B T 2\gamma) \exp(-2Q/k_B T)$, as follows from eqn (2.31). Assuming that the escape rate is smaller than 10^{-2} 1/s we get $13k_B T$ as a lower limit of Q. In reality the two heads cannot be handled as one larger head, because they are not fixed to each other rigidly; therefore, this is a very low limit. $Q \approx 20k_B T$ is expected to be a better estimate.

We can assume that at the beginning of a stepping cycle both heads sit in the same well of the potential. After ATP hydrolysis the strained spring tries to stretch, pushing one of the heads into a neighbouring well. When the load force is small, it has a large probability $p_{0L}^+ = k_{0L}^+/(k_{0L}^+ + k_{0L}^-)$ that the front head jumps to the forward direction, because of the asymmetry of the potential wells, and it has only a small probability $p_{0L}^- = k_{0L}^-/(k_{0L}^+ + k_{0L}^-)$ that the back head jumps backward. k_{0L}^+ and k_{0L}^- denote the corresponding jumping rate constants. Increasing the load the probability of the forward jump decreases while that of the backward jump increases. The average lifetime $t_{0L} = 1/(k_{0L}^+ + k_{0L}^-)$ of the stressed state also slightly increases. Completing this subprocess the second conformational change occurs: the spring tries to contract. Now, for low load force the probability $p_{L0}^+ = k_{L0}^+/(k_{L0}^+ + k_{L0}^-)$ that the backward head jumps forward to the next well, where the other head is sitting, is close to 1, while the probability $p_{L0}^- = k_{L0}^-/(k_{L0}^+ + k_{L0}^-)$ that the forward head jumps backward is very small. An increasing load has a similar effect on these jumping rate constants as in case of stretching. Thus, under low load force the kinesin molecule steps 8 nm forward during almost each mechanochemical cycle. But on increasing the load the probability that the molecule remains at the same place or even takes a backward 8 nm step increases. On reaching the stall load the average displacement of the kinesin becomes zero.

Since the potential wells are deep compared to $k_B T$, the jumping rate constants can be calculated from the Kramers formula (2.31) with a potential that is the sum of the ratchet potential, the quadratic spring potential, and the linear potential of the load force. From the jumping rate constants one can calculate the average displacement, $d = (p_{0L}^+ p_{L0}^+ - p_{0L}^- p_{L0}^-)L = (k_{0L}^+ k_{L0}^+ - k_{0L}^- k_{L0}^-)[1/(k_{0L}^+ + k_{0L}^-)][1/(k_{L0}^+ + k_{L0}^-)]L$, and duration, $t = t_{0L} + t_{L0} = 1/(k_{0L}^+ + k_{0L}^-) + 1/(k_{L0}^+ + k_{L0}^-)$, of a step. On increasing the load the average displacement decreases due to the increasing probabilities of remaining and slipping back, and the average time slightly increases as a manifestation of the Fenn effect [72]. (At stall load it is about three times larger than without load.)

At saturating ATP (Fig. 5.15(a) and (c)) concentration the only rate-limiting factor is the stepping process; therefore dividing the average displacement by the average time gives the average velocity $v = d/t$ of the kinesin. However, at low ATP concentration (Fig. 5.15(b)) the rate-limiting factor is the diffusion of the ATP to the kinesin; thus, the average velocity is proportional to the average displacement during one cycle multiplied by the ATP consumption rate

$$v(c_{ATP}) = \frac{v_{sat} c_{ATP}}{K_m + c_{ATP}} \approx \frac{v_{sat}}{K_m} c_{ATP} \approx \text{const } c_{ATP}, \tag{5.6}$$

where c_{ATP} denotes here the ATP concentration, $v_{sat} = 1/t$ is the inverse of the cycle time, and K_m is the mechanochemical Michaelis–Menten constant. The difference

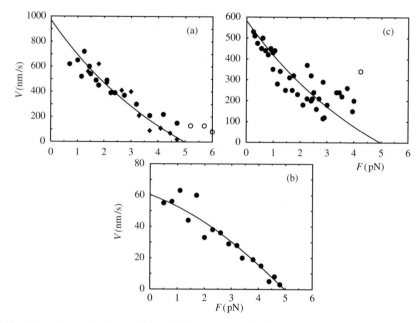

Fig. 5.15 The force–velocity curves for an individual kinesin molecule at saturating ($\gg$ 90 µM) ATP in (a) and (c), and at low ($\sim$10 µM) ATP concentration in (b). The experimental data (scattered symbols) are from Svoboda and Block [14] in (a) and (b) obtained by using optical tweezers, and from Hunt *et al.* [13] in (c) applying viscous load. The open circles correspond to the simultaneous effect of multiple kinesin motors in the authors' interpretation [13, 14]. The circles and diamonds in (a) mean two different set of the measured data. To get the best fit (solid lines) we chose 0.7 nm for the backward length of the potential wells and 1.75 nm for the forward length; the depth of the wells was $Q = 20k_BT$. In plot (b) we multiplied the average displacement by the ATP consumption rate $v(c_{ATP}) = 10$ 1/s for the best fit. In plot (c) we used the same curve as in (a) but multiplied by a factor of 0.6, which accounts for the increased viscosity (after [69]).

between the shapes of the force–velocity curves at saturating and low ATP concentrations shows the Fenn effect unambiguously.

Note that during the entire calculation the concentrations of the ATPase products (ADP and P_i) have been neglected. Though this assumption is supported by the *in vitro* experiments, a thermodynamically complete description of the model cannot be given without taking into account the reverse processes of the ATP hydrolysis.

A typical trajectory of the model with a load force that increases linearly with the displacement of the kinesin (as in optical trapping) can be seen in Fig. 5.16. This curve is very similar to the experimental data [14, 15]. For large forces (i.e., for large displacements) the backward steps can be observed.

Long-Stride model

A Long-Stride model (Fig. 5.17)for the motion of kinesin was introduced by Peskin and Oster [73]. In this model the heads are also connected at a hinge and move along a single protofilament. The heads are initially bound to two nearest β-tubulin sites of the protofilament, separated by 8 nm. It is supposed that ATP hydrolysis by either

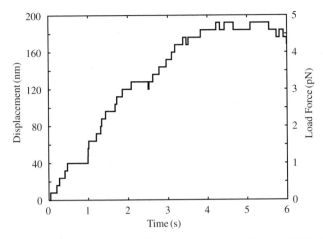

Fig. 5.16 A typical simulated trajectory of kinesin in an optical trap with strength 0.04 pN/nm.

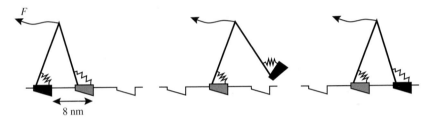

Fig. 5.17 Illustration of the stepping of kinesin in the Long-Stride model.

head weakens its affinity to the microtubule, but after releasing P_i (or ADP) the head recovers its high affinity. The key constituents of the model that drive the motor forward are that (i) the preferred configuration of a bound head is the one in which the head leans forward (the rest length of the spring in Fig. 5.17 is the short one) and (ii) and the hydrolysis rate is greater for the back head than for the front head (e.g., because the strain induced in the front head adversely affects its hydrolysis site, slowing or arresting its catalytic affinity).

Thus, as the ATP hydrolysis in the back head weakens its affinity to the microtubule, the back head detaches, allowing the front head to lean forward into its preferred configuration. The diffusive motion of the detached head is restricted to the interval [-8 nm, 8 nm] with respect to the position of the bound head. After releasing the P_i (or ADP) the free head recovers its high affinity and binds to the first empty β-tubulin site that it encounters. The two sites that are within reach are the ones 8 nm in front and 8 nm in back of the bound head. Binding is biased towards the back site by the applied load force, but towards the front site by the forward-leaning tendency of the bound head.

Choosing reasonable parameters, much of the experimental data can be fitted with this model, too.

Duke and Leibler [74] introduced another version of the Long-Stride model in which the heads are capable of undergoing conformational changes and an elastic element connects each head to the rest of the molecule. They investigated cases where the chemical cycles of the heads were either correlated or uncorrelated and found good agreement with the experimental results.

A general drawback of the Long-Stride models is that during the long 16 nm diffusion of the free head it encounters several *beta*-tubulin sites on the same or neighbouring protofilaments, where the head may bind. This would lead to nonuniform step sizes and sidling motion. The Two-Step models give more stability to the protofilament tracking and yields uniform step sizes.

Chemically reversible model

In the above models the direction of the kinesin's motion is determined either by the asymmetry of the interaction between the head and the microtubule or by the conformation of the heads in different chemical states. Reversing the direction of motion would require a change in these characteristics of the models. As shown by recent experiments [21,22] changing a few amino acids in the neck region of the heads is enough to reverse the directionality of both kinesin and ncd, and it is hard to imagine that these tiny changes would affect the interaction potentials or the conformations so much.

Based on the fact that the neck region is close to and interacts with the ATP binding socket [22] we present here another ratchet model (which belongs to the Two-Step models), where the direction of motion is controlled by subtle differences in the chemical mechanism of ATP hydrolysis, rather than by differences in the structure of the motor or in its interaction with the microtubule [75,76]. Figure 5.18 illustrates the interaction potential between the head and the protofilament, which depends on the chemical state of the head. In the E^{ADP+P_i} and E^{ADP} states of the molecule there are two potential wells within each period: one with a relatively high energy minimum (H) and the other one with a relatively low energy minimum (L). When the motor gets into either of these chemical states, there is first a fast local equilibration in the H well, followed by a slower relaxation from H to L. When the motor is in the E^{ADP+P_i} state, a transition from H to L to the right is allowed, while the transition to the left is blocked by a high energy barrier. When the motor is in the E^{ADP} state, a transition from H to L to the left is allowed, but the transition to the right is blocked.

In this model the direction of motion changes when the relative rates for releasing P_i and ADP change. The solid arrows trace the case for slow P_i and fast ADP release. Because P_i release is slow, the motor in the E^{ADP+P_i} state relaxes with high probability to L to the right before P_i dissociates. Following the release of P_i, the motor rapidly equilibrates in the H well of the E^{ADP} state. But because ADP release is fast, ADP most probably dissociates before relaxation to L can occur. Thus, completing one chemical cycle takes the motor one period to the right.

On the other hand, if P_i is fast and ADP release is slow, the motor moves one period to the left for every cycle (dashed arrows). So simply by changing the chemical

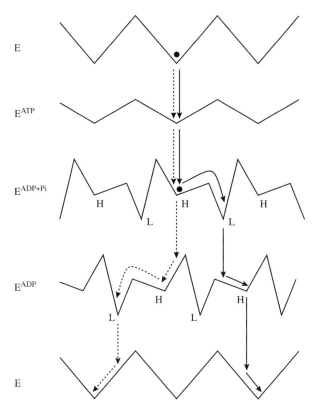

Fig. 5.18 Ratchet mechanism for chemically reversible motion. The motor concomitantly cycles through its chemical states while catalysing ATP hydrolysis (in the vertical direction) and translocates through space along a microtubule (horizontally) (after [76]).

transition rate constants relative to the mechanical ones, one can reverse the direction of the motion.

But there is a second, even simpler way of reversing the directionality, by changing the chemical specificity of the motor in the H and L positions. It is easy to see that if L is specific for P_i release in the E^{ADP+P_i} state and H is specific for ADP release in the E^{ADP} state, the motor moves to the right (solid arrows). Or, if H is specific for P_i release and L is specific for ADP release, the motor moves to the left.

Introducing some cooperativity between the two heads of the molecule (e.g., ATP hydrolysis at one head induces ATP binding to the other, but ATP hydrolysis at this second head cannot proceed until ADP dissociates from the first head) and choosing reasonable parameters, the results of the model are in very good agreement with the experiments. Note that the low energy barriers in the E^{ATP} state are necessary for a Two-Step motion, because it allows a passive 8 nm step of one head while the second head takes an active 8 nm step (by completing its chemical cycle) and tries to drag the first head along.

While in the preceding models the presence of the two heads of kinesin (or ncd) was crucial to the motion, this chemically reversible model predicts that a single-headed molecule should display some processivity. Very recent experiments have indeed confirmed that a single-headed kinesin construct can move processively [77].

5.3.2 Myosin

One of the most important problems in the theory of muscle contraction is to derive the macroscopic mechanical properties of the contraction from the microscopic bio-chemical steps of the cross-bridge interaction. The first and very influential analysis was presented by Huxley [78]. With his original two-state model – where the myosin cross-bridge (head) was regarded as an elastic structure – he was able to reproduce much of the experimental data available at that time. Since then the models have become finer and more complicated to fit the increasing quantity of experimental data [79–86]. The general disadvantages of these models are that (i) they incorporate several functions among the other fitting parameters, and (ii) they can be solved only numerically using computer simulations, and thus the relationship between the microscopic and macroscopic parameters cannot be seen clearly.

Here we present a simple four-state model for the cross-bridge interaction which is free from the above-mentioned shortcomings. The model (i) involves the *relevant* kinetic data and other microscopic parameters and (ii) provides simple *analytic* solutions for the mechanical properties of muscle contraction. A similar model was introduced by Duke [87].

The well known empirical Hill equation [88]

$$(F + a)(V + b) = c \tag{5.7}$$

also gives a very good fit to the measured force–velocity relationship and has been used so far, although, its relationship with the biochemical level is unknown. F and V denote the force and velocity of the contractive muscle, respectively, and a, b, c are the fitting parameters of the equation.

Myosin cross-bridge model

In the present model we follow the framework of the original approach of Huxley [78] but adapted to the recent scheme of cross-bridge cycle mentioned in §5.1.2. To be able to handle the model analytically we have to make some careful simplifications that have only an insignificant effect on the final results. The model is based on the following assumptions:

1. Let us suppose that the distance between two neighbouring binding sites on the actin filament – having good orientation according to the myosin heads on the myosin filament – is $d \approx 37$ nm, being the half-period of the double helix of the actin filament. The position of the actin filament is fixed while the myosin

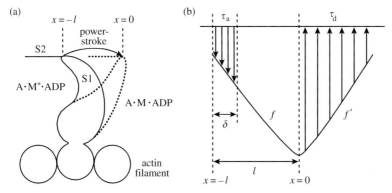

Fig. 5.19 Cartoon (a) shows the conformational change in the myosin head during the power-stroke. In the schematic picture (b) the flat and 'V'-shaped potentials represent the weakly and strongly bound states of the myosin cross-bridge. The slopes of the 'V'-shaped potential are f and f', corresponding to the force produced by the myosin head in the power-stroke and in the drag-stroke region, respectively. Transition from the weakly bound state to the strongly bound state can occur in the region $-l < x < -l + \delta$ with characteristic time τ_a and transition from the strongly bound state to the weakly bound state in the region $0 < x$ with characteristic time τ_d (after [47]).

filament passes by it with a velocity V from left to right. Since the spacing of the myosin heads and the actin binding sites are incommensurate, the distances between the myosin cross-bridges and their nearest binding sites are uniformly distributed within the range $[-d/2, d/2]$ at any time. Thus, the macroscopic properties of muscle contraction can be calculated as a spatial average for the position of one cross-bridge over this interval or, which is equivalent, a temporal average over the period d/V during which the cross-bridge traverses the interval. Therefore, in the following we will consider the motion of one single myosin cross-bridge.

2. To set the notations let x denote the longitudinal position of the cross-bridge relative to the nearest binding site $(-d/2 \leq x \leq d/2)$. More precisely, let x denote the position of the junction point of the S1 and S2 subfragments [see Fig. 5.19(a)]. x is taken as 0 nm when the A · M · ADP state is unstrained.

3. Initially the myosin cross-bridge is in a weakly bound, non-force-producing (A) · M · ADP · P_i state, which can be represented by a flat potential (Fig. 5.19(b)).

4. When the cross-bridge gets into the vicinity $-l \leq x \leq -l + \delta$ of the actin binding site, phosphate can be released with a rate constant k_{P-} forming a stereospecific, strongly bound, force-producing A · M* · ADP state. The reverse reaction is characterized by the second-order rate constant k_{P+}. (The − sign in the subscript always refers to dissociation and implies that the rate constant is first-order, and the + sign refers to association with a second-order rate constant.) Leaving this interval without forming the strongly bound state, the cross-bridge has no more chance to bind strongly to this actin binding site, and remains in the weakly bound (A) · M · ADP · P_i state. The width δ of this interval is the composition

of the Debye length and the rotational freedom of the myosin head, and can be well approximated by a few nanometers. In the following we will use $\delta = 3$ nm.

5. The notation ' * ' in the $A \cdot M^* \cdot ADP$ state means that the cross-bridge is already in the force-producing $A \cdot M \cdot ADP$ state, but it still stays within the interval δ and has some chance to detach by phosphate binding. The force produced by these states in the power-stroke region ($-l \leq x \leq 0$) is about $f \approx 4$ pN, and the power-stroke working distance is $l \approx 11$ nm (see §5.1.2). Finishing the power-stroke continued attachment resists useful work, i.e., exhibits a negative force $-f'$ in this *drag-stroke* region ($0 < x$). f' must be close to f, but not necessarily equal to it. Altogether, the strongly bound $A \cdot M \cdot ADP$ state – including the $A \cdot M^* \cdot ADP$ state – can be represented by a 'V'-shaped potential shown in Fig. 5.19(b).

6. In the drag-stroke region ($0 < x$), to protect the cross-bridge from the negative force, ADP is released with a rate constant k_{ADP-} followed by a transition to the still strongly bound $A \cdot M$ state, and then ATP can bind with a second-order rate constant k_{ATP+}, resulting in rapid cross-bridge detachment. The reverse transition $A \cdot M$ to $A \cdot M \cdot ADP$ has the second-order rate constant k_{ADP+}, but the reverse transition $(A) \cdot M \cdot ATP$ to $A \cdot M$ can be neglected. The strongly bound $A \cdot M$ state is supposed to have similar structural properties to the $A \cdot M \cdot ADP$ state; therefore, it can be represented by the same 'V'-shaped potential mentioned above.

7. Finally, ATP hydrolysis takes the cross-bridge from the weakly bound $(A) \cdot M \cdot ATP$ state to the also weakly bound $(A) \cdot M \cdot ADP \cdot P_i$ state, which is free to participate in a new working cycle. Since these weakly bound states generate basically no force, they can be represented by a flat potential.

Unlike in all the other models here we use a 'V'-shaped piecewise linear potential instead of a parabolic-shaped potential of a linearly elastic cross-bridge. We have performed the calculations with both kinds of potentials, but it has turned out that the results fit the experimental data better if we use the 'V'-shaped potential. In addition, the analytic results are much simpler in this case. Moreover, the shape of the potential is essential only in the drag-stroke region, because in the $[-l + \delta, 0]$ part of the power-stroke region the strongly bound heads are uniformly distributed, thus the force exerted by them can always be substituted by their average force over that interval.

The rate constants introduced above should be checked for consistency with thermodynamics. Since the strongly bound $A \cdot M \cdot ADP$ and $A \cdot M$ states have similar mechanical properties corresponding to the same 'V'-shaped potential, their Gibbs free energy can differ only by a constant value, implying that the ratio k_{ADP-}/k_{ADP+} (where they are not zero) must be independent of x. Because both rate constants are independent of x in the model, this requirement is trivially fulfilled. The rate k_{ATP+} can be chosen arbitrarily, because its reverse rate is neglected. In the model the rates k_{P-} and k_{P+} (where they are not zero) are also independent of x, though the

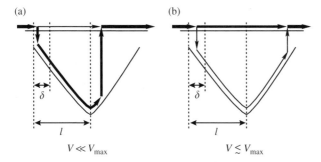

Fig. 5.20 The two pictures show how the typical myosin head passes by the actin binding site for small and large velocities. The thickness of the arrows indicates the probability of the corresponding state.

$(A) \cdot M \cdot ADP \cdot P_i$ and the $A \cdot M \cdot ADP$ states correspond to a flat (x-independent) and a 'V'-shaped (x-dependent) potential, respectively. Since the rates are non-zero only in the small δ region, the potential difference does not change too much there, and these rates can be considered to be constant. Or, in other words, within this small δ region the rates are approximated with their average. Thus, the model is a reasonable reduction of a thermodynamically consistent system.

Now let us see how the model works. At low speed (Fig. 5.20(a)) when the cross-bridge reaches the vicinity of the binding site, the ratio of the strongly and weakly bound states reaches its stationary value already at the beginning of the δ region. If the phosphate concentration is not too high, a large portion of the cross-bridges gets into the strongly bound state. These cross-bridges produce the force f in the whole power-stroke region, then detach at the beginning of the drag-stroke region.

On the other hand, at high speed (Fig. 5.20(b)) only a smaller proportion of cross-bridges has the chance to get into the strongly bound state, and then at the end of the power-stroke these cross-bridges are carried deeply into the drag-stroke region. At the maximum velocity the net force is zero, because the positive and negative forces are balanced.

Thus, increasing the velocity, the average force produced by the myosin cross-bridges decreases, because (i) the portion of the strongly bound states decreases, and (ii) the distance covered in the drag-stroke region increases. On reaching the maximum velocity the force drops to zero.

From the mechanical point of view it is easy to see that the transition from the weakly bound $(A) \cdot M \cdot ADP \cdot P_i$ state to the strongly bound $A \cdot M^* \cdot ADP$ state can be simply characterized by an 'attachment' time constant τ_a and a weight w, as a consequence of the constant force f in the δ region. (The weight w is the steady-state probability of the strongly bound state which is approached exponentially with the time constant τ_a.) Similarly, solving the corresponding kinetic equations in the drag-stroke region, it turns out that the transition from the strongly bound $A \cdot M \cdot ADP$ state to the weakly bound $(A) \cdot M \cdot ATP$ state – through the strongly bound $A \cdot M$ state – can be characterized by a 'detachment' time constant τ_d (see Fig. 5.19(b)).

The values of these parameters are as follow:

$$\tau_a = \frac{1}{k_{P-} + k_{P+}[P_i]}, \qquad w = \frac{k_{P-}}{k_{P-} + k_{P+}[P_i]},$$

$$\tau_d = \frac{1}{k_{ADP-}} + \frac{1}{k_{ATP+}[ATP]} + \frac{k_{ADP+}[ADP]}{k_{ADP-} \cdot k_{ATP+}[ATP]}. \qquad (5.8)$$

The probability that the cross-bridge gets into the force-producing state within the δ long region is

$$p_a = w\,(1 - \exp(-\delta/V\tau_a)). \qquad (5.9)$$

The average distances covered by an attached cross-bridge in the positive and negative force regions are

$$l_+ = l - \delta\left(\frac{V\tau_a}{\delta} - \frac{1}{\exp(\delta/V\tau_a) - 1}\right) \quad \text{and} \quad l_- = V\tau_d, \qquad (5.10)$$

respectively. The second term in l_+, which is a straightforward result of the corresponding kinetic equation, is zero at zero velocity and converges to $\delta/2$ for increasing velocities, as one would expect.

Using the notation

$$V_a = \frac{\delta}{\tau_a} \quad \text{and} \quad V_d = \frac{l}{\tau_d}$$

one can easily calculate the average force F produced by a cross-bridge and the duty ratio R (which is the fraction of the attached cross-bridges) from eqns (5.9) and (5.10) by averaging over the interval $[-d/2, d/2]$: $F = p_a(fl_+ - f'l_-)/d$ and $R = p_a(l_+ + l_-)/d$, i.e.,

$$F = \frac{l}{d} f w \left(1 - \exp(-V_a/V)\right)\left[1 - \frac{\delta}{l}\left(\frac{V}{V_a} - \frac{1}{\exp(V_a/V) - 1}\right) - \frac{f'}{f}\frac{V}{V_d}\right], \qquad (5.11)$$

$$R = \frac{l}{d} w \left(1 - \exp(-V_a/V)\right)\left[1 - \frac{\delta}{l}\left(\frac{V}{V_a} - \frac{1}{\exp(V_a/V) - 1}\right) + \frac{V}{V_d}\right]. \qquad (5.12)$$

The actual dependence of the force F on the velocity V as given by this simple analytic formula is very similar to the hyperbola of the Hill equation (5.7) (see Fig. 5.21). The slight deviation at low velocities was experimentally observed [89]; however, it is usually not much greater than the experimental error. The similarity between the two curves comes from the term $1 - \exp(-V_a/V)$ which originates from the attachment probability p_a. The eqns (5.11) and (5.12), just like the Hill equation, have 3 free parameters: V_a, $V_d f/f'$ and the coefficient $f w l/d$. The fourth parameter δ/l is essentially fixed and its value is around 0.3. Since the positive and negative forces

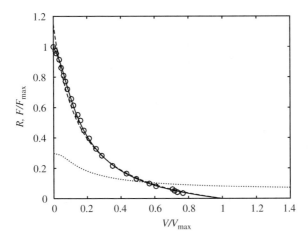

Fig. 5.21 The open circles show the experimentally measured force–velocity relationship of frog semitendinosus muscle [89] and the solid line is the fit of the model. The dashed line is a fit according to the Hill equation which has a slight deviation from the experimental data at low velocities. The dotted line is the duty ratio predicted by the model.

produced by the cross-bridges must be similar to each other, their fraction f'/f is approximately 1.

The maximum force produced by one myosin cross-bridge on average can be achieved at zero velocity. Then eqn (5.11) gives

$$F_{max} = \frac{l}{d} f w. \tag{5.13}$$

Since the typical phosphate concentration is low in normal muscle tissues, the weight w is close to 1 and the maximum force is approximately $F_{max} \approx 1.2\,\text{pN}$, which is consistent with the experiments.

The maximum velocity can be determined from the condition that the average force is zero, i.e., the last factor of eqn (5.11) is zero. Although no exact analytic solution can be obtained, a lower and upper bound can be given. Since the second term in the square brackets is within the range $[0, \delta/2l]$,

$$\left(1 - \frac{1}{2}\frac{\delta}{l}\right)\frac{f}{f'}\frac{l}{\tau_d} \leq V_{max} \leq \frac{f}{f'}\frac{l}{\tau_d}. \tag{5.14}$$

These two bounds are close to each other and their difference is only about 15%. Therefore, the velocity

$$V_0 = \frac{f}{f'}\frac{l}{\tau_d} \tag{5.15}$$

provides a good estimate of the maximum. Note that in typical cases V_{max} is several times larger than V_a. Therefore, the lower bound is a better approximation for the maximum velocity.

Introducing the new notations

$$F^* = \frac{F}{F_{max}}, \quad V^* = \frac{V}{V_0}, \quad C = \frac{V_a}{V_0} = \frac{f'}{f} \frac{\delta}{l} \frac{\tau_d}{\tau_a}$$

eqn (5.11) can be written as

$$F^* = (1 - \exp(-C/V^*)) \left[1 - \frac{\delta}{l} \left(\frac{V^*}{C} - \frac{1}{\exp(C/V^*) - 1} \right) - V^* \right]. \tag{5.16}$$

Here the meaning of the three parameters is very simple: F_{max} is the maximum average force, V_0 is roughly the maximum velocity and C is in close relation to the curvature of the force–velocity curve. For example, after determining the maximum force and velocity in Fig. 5.21 the only fitting parameter was $C = 0.15$. The great advantage of this model is that these important mechanical parameters can be clearly expressed in terms of the microscopic parameters.

The measurement of the duty ratio R is very difficult, and the experimental results are contradictory. This model predicts (see eqn (5.12) and Fig. 5.21) that it starts from $w\delta/l \approx 0.3$ at zero velocity (isometric contraction) and that on reaching the maximum velocity it drops to only a several percent, for typical values of the parameters.

From the model one can also determine the ATPase rate v_{ATPase} of one myosin cross-bridge. Because one ATP is used in each cycle, the rate can be calculated as the reciprocal value of the time V/d (needed to advance the distance d multiplied by the cross-bridge attachment probability p_a:

$$v_{ATPase} = \frac{V}{d} w(1 - \exp(-V_a/V)). \tag{5.17}$$

During the entire calculation we have ignored the ATP hydrolysis step and we have supposed that each cross-bridge that reaches the δ region is already in the $(A) \cdot M \cdot ADP \cdot P_i$ state. Since the hydrolysis step itself is a fast process (with a rate $\simeq 100$ 1/s) with a large equilibrium constant ($\simeq 10$), this is indeed a good approximation. In the low speed regime it is obvious that after the detachment the cross-bridge has enough time to hydrolyse the ATP before reaching the vicinity of the next binding site. For large velocities this time is not enough, but because the phosphate release rate is smaller than the ATP hydrolysis rate (see later) and phosphate can be released only in the δ region while hydrolysis can occur anywhere, the attachment is limited only by the phosphate release, and the ATP hydrolysis step can be neglected.

Now let us see some examples of how the various experimental results can be interpreted and clarified by this model. There is much experimental evidence (e.g., [90–93]) to show that on increasing the ATP concentration the maximum shortening velocity V_{max} exhibits a classical Michaelis–Menten (M–M) saturation behaviour:

$$V_{max} = V_{max}^{\infty} \frac{[ATP]}{K_m + [ATP]}, \tag{5.18}$$

where V_{max}^∞ represents the maximum shortening velocity at infinite ATP concentration, and K_m is the M–M constant. In addition, Cooke and Pate [91] showed that ADP acts as a pure competitive inhibitor of V_{max} by increasing K_m:

$$K_m = K_m^0(1 + [ADP]/K_i), \tag{5.19}$$

where K_m^0 is the M–M constant in the absence of ADP, and K_i is the inhibition constant.

The quantity V_0 of the model exactly shows this simple ATP and ADP dependence. Consequently, from eqns (5.8) and (5.15) one can easily express the V_{max}^∞, K_m^0, and K_i (macroscopic) constants from the (microscopic) kinetic constants:

$$V_{max}^\infty = \frac{f}{f'}lk_{ADP-}, \quad K_m^0 = \frac{k_{ADP-}}{k_{ATP+}}, \quad K_i = \frac{k_{ADP-}}{k_{ADP+}}. \tag{5.20}$$

For typical ATP concentrations the maximum shortening velocity is $1 - \delta/2l \approx 0.85$ times smaller than V_0, and only for low concentrations are V_{max} and V_0 equal. Thus, the maximum shortening velocity also exhibits the observed M–M saturation behaviour with only a small, less than 15%, deviation at low ATP concentrations.

Figure 5.22(a) and (b) show the force–velocity relationship of rabbit fast-twitch psoas muscle and rabbit slow-twitch semimembranosus muscle, respectively, for various ATP concentrations. The scattered symbols are experimental data [90, 93] and the solid lines are fits to the model. For psoas muscle we used the following fitting parameters: $k_{ADP-} = 230$ 1/s, $k_{ATP+} = 0.76$ 1/μM s, and $k_{P-} = 76$ 1/s, and for semimembranosus $k_{ADP-} = 76$ 1/s, $k_{ATP+} = 3.8$ 1/μM s, and $k_{P-} = 25$ 1/s. Thus the M–M constants are about 300 and 20 μM, respectively. Similarly good fit can be achieved for varying the ADP concentrations as well. The ratio f'/f was set to 1 in each fit.

Note that the isometric contraction is a singular point of the model, because for small positive velocities the interval $[-l, 0]$ is filled by the cross-bridges uniformly, but for small negative velocities the interval $[-l + \delta, 0]$ is empty and the strongly bound cross-bridges can get left from the $x = -l$ position. This discontinuity at zero velocity had already been experimentally observed by Katz [94]. Therefore, during the fitting we did not use the measured isometric force.

From eqn (5.13) it follows that the isometric force does not depend on either the ATP or the ADP concentration. This is in good agreement with the experiments [90, 93] that report only a slight change in the isometric force. This change can be explained by small non-zero ADP release and ATP binding rates in the power-stroke region.

Conversely, increasing phosphate concentration [P_i] has been experimentally shown to decrease the isometric force while slightly increasing the maximum shortening velocity [91]. In addition, the effect of P_i is independent of the effect of ATP and ADP. These observations are also consistent with the model. From eqn (5.8) it follows that the P_i concentration has an effect only on the weight w and the time constant τ_a, while the ATP and ADP concentrations have an effect on the time constant τ_d. Thus, on increasing the P_i concentration the isometric force goes to zero as a hyperbola and

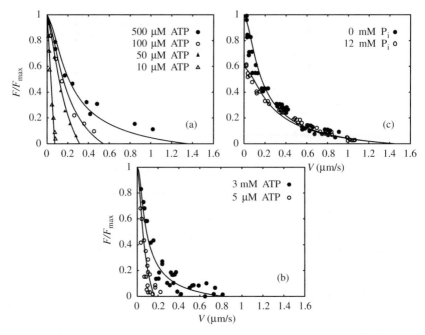

Fig. 5.22 The force–velocity relationship of the contracting muscle in various situations: rabbit psoas muscle [90] in (a) and rabbit semimembranosus muscle [93] in (b) for different ATP concentrations in the absence of ADP and P_i; and rabbit psoas muscle [91] in (c) for two different P_i concentrations in the absence of ADP at saturating (4 mM) ATP. The scattered symbols are experimental data and the solid lines are fits of the model (after [47]).

the parameter C goes to infinity linearly, causing the curvature of the force–velocity curve also go to zero. The characteristic velocity V_0 does not change, however, it can be shown that the maximum velocity slightly increases.

Figure 5.22(c) shows the force–velocity relationship of rabbit psoas muscle in the absence of P_i and in the presence of 12 mM P_i. The scattered symbols are experimental data [91]. To achieve the best fit (solid lines) we used the following fitting parameters: $k_{ADP-} = 155$ 1/s, $k_{P-} = 56$ 1/s, and $k_{P+} = 3.1$ 1/mM s. The deviation of the common fitting parameters of the two different experiments in Fig. 5.22(a) and (c) are within a factor of 1.5.

The advantage of the 'V'-shaped potential can be seen in Fig. 5.22(a) and (b). For low ATP concentrations the velocity is always small, thus eqn (5.16) can be approximated by $F^* = 1 - V^*$. For a parabolic-shaped potential this would be $F^* = 1 - V^{*2}$, which is a concave function. The experiments show that for decreasing ATP concentration the curvature of the force–velocity relationship decreases but never changes its sign.

The ATPase activity v_{ATPase} as a function of the ATP concentration also exhibits a classical M–M saturation behaviour [95], but with an order of magnitude smaller M–M constant $K^0_{m,ATPase}$ than K^0_m. The model does not give an exact M–M formula for

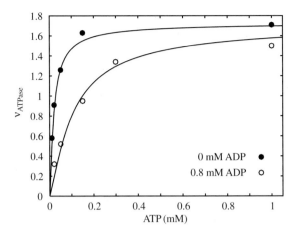

Fig. 5.23 The ATPase rate v_{ATPase} of rabbit psoas muscle as a function of the ATP concentration for two ADP concentrations in the absence of P_i. Scattered symbols are experimental data [95] and solid lines are fits of the model.

the ATPase activity, but, surprisingly, determining the maximum shortening velocity from eqn (5.11) and then substituting it into eqn (5.17) results in a curve which is very similar to a hyperbola.

Figure 5.23 shows the ATPase rate of rabbit psoas muscle for zero and 0.8 mM ADP concentrations. Fitting the experimental data [95] we obtained $k_{ADP-} = 60$ 1/s, $k_{ADP+} = 0.35$ 1/μM s, $k_{ATP+} = 0.2$ 1/μM s, and $k_{P-} = 22$ 1/s. These rate constants are 3–4 and 2–3 times smaller than the rate constants of Fig. 5.22(a) and (c), respectively. However, their ratios, the M–M constant $K_m^0 = 300$ μM and the inhibition constant $K_i = 170$ μM, are the same and agree with other experiments. The discrepancy may come from the fact that the ATPase activity v_{ATPase} of one myosin cross-bridge is very difficult to measure because it presumes the exact knowledge of the number of myosin heads that potentially participate in the muscle contraction. Supposing that only one of the two heads of a myosin molecule can work at a time results in a 2 times higher ATPase rate, thus a factor of 2 can be easily gained. The M–M constant of the ATPase activity is $K_{m,ATPase}^0 = 20$ μM, which is, indeed, about an order of magnitude smaller than K_m^0. The model gives a similarly good fit to the ADP and P_i dependence of the ATPase activity.

An interesting property of the model is that for large ATP concentrations, when $V \gg V_a$, the ATPase rate is $v_{ATPase} \approx wV_a/d \approx k_{P-}\delta/d$, which is independent of both the ADP and the P_i concentrations.

The model can also explain the first part of the mechanical transient resulting after rapid change in muscle length or tension [96]. The second part seems to be the result of a more complex process, and requires the extension of the model.

Some very recent experiments on individual myosin molecules also require further extensions of the model: an optical tweezers experiment combined with fluorescence microscopy indicates that besides the widely accepted scheme of Lymn and Taylor

(§5.1.2) the cross-bridge cycle may follow alternative kinetic schemes as well [97]. Another experiment with optical tweezers shows that certain types of myosin produce the power-stroke in two steps [98]. A scanning probe experiment by Yanagida's group [99] has led to a very peculiar result, which does not fit into any proposed model or mechanism of the cross-bridge cycle: their data indicate that a single myosin molecule is able to take several steps of about 5.3 nm (the distance between adjacent actin monomers) for the hydrolysis of only one ATP.

In conclusion, this theoretical model directly connects the experimentally observed macroscopic behaviour with the microscopic features of the cross-bridge cycle, leading to a deeper understanding of the mechanisms of muscle contraction. In particular, an analytic formula for the force–velocity relationship containing the kinetic data is very useful in estimating the relative weight of the various factors and subprocesses during muscle contraction. An important consequence of the results is that for a theoretical description of muscle contraction it is sufficient to use rather simple assumptions for the actual forms of the underlying interactions (potentials, rate constants), indicating that many of the details previously considered are not necessary to obtain predictions in good agreement with the experiments. This is fully consistent with the picture emerging from various studies in statistical physics, where it has theoretically been demonstrated that in most systems with many identical units the overall behaviour of the systems only weakly depends on the details of the simultaneously acting units.

5.3.3 ATP synthase

The qualitative mechanism of the operation of the ATP synthase is thought to be known (see §5.1.3). Based on the recent experimental results here we present a simple quantitative mechanochemical description for both the F_0 and the F_1 component of the enzyme, shedding light on the details of its operation [100]. Similar models have been proposed by the group of Oster [101–103].

The description is based on the following general assumptions: (i) each chemical state, i, of either component determines a free-energy profile (FEP), $G_i(\phi)$, in which the shaft moves (rotates) with Brownian dynamics; and (ii) the transition rate constants between the chemical states (denoted by $k_{i,j}(\phi)$ from state i to j) depend on the (angular) position, ϕ, of the shaft, and these rate constants must, of course, obey the relation $k_{i,j}(\phi)/k_{j,i}(\phi) = \exp[(G_i(\phi) - G_j(\phi))/(k_B T)]$ for thermodynamic consistency.

Mechanochemical model of the F₁ component

Since each catalytic unit of the F_1 component has at least $n = 3$ different chemical states: empty, ATP-bound, and ADP+P_i-bound ($n = 4$ or 5, if ADP and P_i can dissociate separately), the number of possible FEPs associated with the F_1 component is quite big: n^3. Though this number can be reduced by considering the threefold rotational symmetry of the α-β-ring, it remains still too big to perform a reliable analysis. However, supposing that the interaction between the γ-shaft and a β-subunit is *independent* of the chemical state of the other two β-subunits, each of the n^3 FEPs can be generated from the set of only n primary FEPs, corresponding to the interaction between the γ-shaft and the n chemical states of a β-subunit. Further, we assume that the FEPs are *smooth* enough that the interaction can be sufficiently characterized by the interaction energies at three discrete relative positions of the γ-shaft and the β-subunit (separated by $120°$). Then the interaction energies elsewhere can be interpolated by connecting these three points with a smooth curve. Denoting the $n = 3$ states of the β-subunit by 'ATP', 'ADP + P$_i$', and 'Empty' and marking the three sides of the γ-shaft (that were facing the ATP-containing, ADP+P$_i$-containing, and empty β-subunits in the crystallized structure) by 'T', 'D', and 'E', the interaction energies are

$$T\begin{cases} \text{ATP} & - & 0 \\ \text{ADP+P}_i & - & E_{TD} \\ \text{Empty} & - & E_{TE} - \Delta G^*_{ATP} \end{cases}$$

$$D\begin{cases} \text{ATP} & - & E_{DT} \\ \text{ADP+P}_i & - & 0 \\ \text{Empty} & - & E_{DE} \end{cases} \qquad E\begin{cases} \text{ATP} & - & E_{ET} + \Delta G^*_{ATP} \\ \text{ADP+P}_i & - & E_{ED} \\ \text{Empty} & - & 0 \end{cases}$$

Without loss of generality one of the $n = 3$ interaction energies for each side of the shaft can be chosen arbitrarily and is set to zero. The free energy of ADP + P$_i$ in the solution can also be set to zero. Then the free energy of ATP must be set to the free energy liberated by an ATP hydrolysis (ΔG_{ATP}). The notation ΔG^*_{ATP}, which is the free energy of the ATP in physiological conditions (usually not more than $24\,k_BT$), is introduced to make the parameter choice more practical. Thus, the FEP for any chemical state of the β-subunits (and for any ATP content of the solution) at the three discrete positions of the γ-shaft can be determined as the sum of the corresponding interaction energies, decreased (increased) by ΔG_{ATP} as many times as ATP was adsorbed from (desorbed to) the solution. Smoothly connecting these three discrete values (as illustrated in Fig. 5.25) results in the complete FEP for each state of the F_1 component. It will turn out later that the smoothness of the FEP (at least in the relevant interval) is crucial to the function.

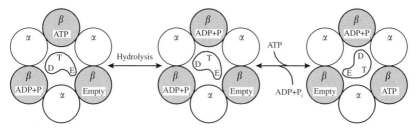

Fig. 5.24 Illustration of the binding change mechanism in the F_1 ATPase. Chemical transitions and the rotation of the γ subunit occur concomitantly in a well-defined order.

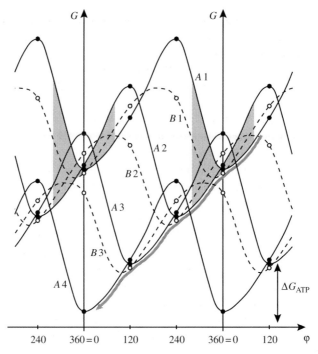

Fig. 5.25 The FEPs for the F_1 component. The γ shaft moves (rotates) along the effective potential (grey arrow) while visiting the FEPs $A1$, $B1$, $A2$, $B2$, ..., subsequently. The circles indicate the free energy values defined in the model, and the A and B curves connect them smoothly.

Now let us take a closer look at the binding change mechanism [34], and see what special constraints are imposed on the model by some of the experimental results. Since in the crystallization process a nonhydrolysable ATP analogue was used, the operation of the F_1 ATPase was blocked right before the hydrolysis in the ATP-containing β-subunit. This implies that the other two β-subunits (the empty and the ADP $+$ P$_i$-containing) got into their energetically most favourable chemical state, and the γ-shaft found its minimal energy position. After this stage, if real ATP were used, the ATP-containing β-subunit would get into its lowest energy state by hydrolysing its ATP. Since the operation of the F_1 ATPase cannot stop here, this transition must modify the FEP such that the shaft can rotate further, as illustrated in Fig. 5.24. But since in this new position the other two β-subunits cannot be longer in their lowest energy state, ADP $+$ P$_i$ release and ATP binding must occur, inducing further rotation of the γ-shaft. Now the F_1 ATPase is in a state which is equivalent to the starting state (but rotated by 120°) and ready to hydrolyse the new ATP.

This binding change mechanism is implemented in Fig. 5.25 for a specific choice of the parameters. The FEP $A1$ corresponds to the chemical state of the F_1 ATPase illustrated by the left drawing of Fig. 5.24. It has a minimum at $\phi = 0$ (where 'T' faces the ATP-containing β-subunit). As a consequence of the ATP hydrolysis the FEP changes to $B1$, which has a lower free energy at $\phi = 0$, but its minimum is shifted to the left, inducing some rotation of the shaft (see the middle drawing of Fig. 5.24). After this rotation the release of ADP + P_i and the binding of ATP changes the FEP to $A2$, which is equivalent to $A1$ but shifted to the left by 120° and lowered by ΔG_{ATP}. Repeating this procedure the FEP changes to $B2$, $A3$, $B3$, ..., successively. The 'effective potential' along which the shaft moves is indicated by a thick grey arrow.

A similar effective potential can be determined for the F_0 motor (discussed later; see Fig. 5.26), and the entire shaft of the ATP synthase (the γ-subunit of F_1 and the c-subunits of F_0 together) moves in the sum of these two effective potentials. The smaller the steepness of this overall potential, the bigger the efficiency of the energy conversion. And since the angular diffusion coefficient of the shaft is quite small ($\sim$ 100–1000 s^{-1} because of the high viscosity of the membrane) a big speed at high efficiency can be achieved only if the relevant parts of the FEPs (along which the shaft moves) have a basically constant slope of $\sim \Delta G_{ATP}^*/120°$. This *maximum power principle* imposes strong constraints on the model. Another indication for the straightness of (the relevant parts of) the potential is that in the motility experiment [39] the rotation of the actin filament was remarkably uniform, and the mechanical work done during one turn was about 3 ΔG_{ATP}^*. Since the angular diffusion coefficients of the actin filaments were very small (< 1 s^{-1} for the longest ones), any considerable (higher than $k_B T$) energy barrier in the path of the shaft would have stopped its rotation for a long period of time.

Now let us estimate the values of the interaction energies of the F_1 ATPase, supposing that $\Delta G_{ATP}^* = 24\, k_B T$. Because at the minimum of $A1$ an ATP hydrolysis step occurs, E_{TD} must be negative and E_{DE}, E_{ET} positive, but with not too large absolute values if we want to locate the transitions near $\phi = 0$ and to achieve a fairly straight $A1$ profile with a basically constant $\Delta G_{ATP}^*/120°$ slope. In Fig. 5.25 we have chosen $-2\, k_B T$ for E_{TD}, and $2\, k_B T$ for E_{DE} and E_{ET}. To allow enough room for the transition from $A2$ to $A1$ (through $B1$) in the reverse direction (i.e., to let $A2$ continue increasing slightly faster than $A1$ when ϕ goes from 0 to a few times 10°), E_{TE}, E_{DT}, and E_{ED} must be large, and are set to 30 $k_B T$ in Fig. 5.25.

We cannot specify the transition rate constants yet. But the transitions leading from $A1$ to $A2$ or back should be fast enough (with rate constants larger than 10^3 s^{-1} or 10^6 s^{-1}M^{-1}) in certain regions of the relevant interval (indicated by the grey area) to allow a high turnover rate (the maximum is about 600 s^{-1}). Other parts of the FEPs (outside the grey area) are never visited by the system and they are irrelevant to the function of the enzyme.

Note that this model is in full agreement with the results of the motility experiment [39]. There the ADP and P_i concentrations were far below their physiological values, yielding a much larger ATP free energy (ΔG_{ATP}) than the physiological value (ΔG_{ATP}^*). In our description this means that the distance (along the G axis) between two subsequent 'A' curves was much larger than in Fig. 5.25. The question is whether the generated torque is closer to $\Delta G_{ATP}^*/120°$ or to the much larger $\Delta G_{ATP}/120°$. The model suggests that it is close to $\Delta G_{ATP}^*/120°$, because torque (force) can be generated only by the slopes of the FEPs (which are independent of the substrate concentration) and never by the chemical transitions. However, a slight increase compared to $\Delta G^*_{ATP}/120°$ is expected, because the transition from $A1$ to $A2$ is irreversible in the absence of ADP and P_i, and the system has a chance to experience the slightly steeper slope of the $A2$ profile between $0°$ and $60°$ without getting back to $A1$. This prediction agrees with the motility experiment, and eliminates the possibility that the measured torque, which was slightly larger than $\Delta G^*_{ATP}/120°$, is the consequence of some slippage, due to the hydrolysis of more than 3 ATPs per revolution.

The very recent motility experiments of Yasuda *et al.* [104] have revealed that the γ-shaft makes discrete $120°$ steps at low ATP concentrations, and in each step it does mechanical work of about ΔG^*_{ATP}, in perfect agreement with the model. Occasional backward steps, with the same speed as that of the forward steps, were also observed. These can be explained either by ATP binding to the wrong sites (such undesirable transitions and the corresponding FEPs should also be included in a more detailed model) or by rare, thermally activated, 'uphill' rotations of the shaft (which, on average, take the same time as 'downhill' rotations).

Mechanochemical model of the F₀ component

The F_0 component of the ATP synthase can be treated similarly to the F_1 component, and an even simpler description can be given with fewer and less complicated FEPs. In the following we consider only those two configurations in which a maximum of one acidic residue of the two c-subunits facing the a-subunit is neutralized by a proton, and we neglect the very improbable configuration in which both acidic residues are protonated at the same time. Besides, this latest configuration would have no significant effect on the function of the ATP synthase if in this situation the motion of the shaft were blocked by a large energy barrier (e.g., due to a positive charge between the two proton ports on the a-subunit, close to the c-subunits), but without a barrier it would decrease the efficiency of the energy conversion by allowing free rotation of the shaft leading to ATP hydrolysis in the F_1 component, decoupled from the proton flow. Since the efficiency of the energy conversion by the ATP synthase is thought to be very good, the omission of this configuration is reasonable.

Because an unprotonated c-subunit cannot enter the membrane (or more precisely, it would cost too much energy), the FEP for the configuration in which both c-subunits are unprotonated is a deep narrow potential well (see curves $B1, B2, \ldots$, in Fig. 5.26), while that for the configuration in which only one of the c-subunits is unprotonated is broader and spans an approximately $360°/N_c$-wide interval ($A1, A2, \ldots$, in Fig. 5.26). And again, from the maximum power principle it follows that in this latest configuration the major part of the FEP (where motion and force generation occur) has a constant slope similar to the that for the F_1 component ($\Delta G^*_{ATP}/120°$).

Thus, the shaft of the F_0 component rotates in the following way (in accordance with [37]). When both c-subunits are unprotonated the shaft is unable to move and sits at the bottom of a narrow potential well (e.g., $B1$). By the protonation of the c-subunit connected to the P-side of the membrane the FEP changes to $A2$, forcing

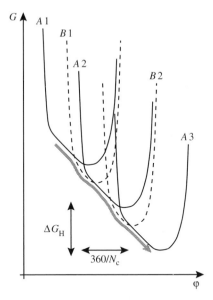

Fig. 5.26 The FEPs for the F_0 component. As in Fig. 5.25 the γ shaft moves along the effective potential (grey arrow) while visiting the FEPs $A1$, $B1$, $A2$, $B2$, ..., subsequently.

the shaft to rotate to the right by one c-subunit. After releasing another proton through the port connected to the N-side of the membrane the FEP changes to $B2$, which is equivalent to $B1$ but shifted to the right by $360°/N_c$ and lowered by the free-energy difference between the two sides of the membrane for a proton (ΔG_H). However, by the protonation of the c-subunit connected to the N-side the FEP can change from $B1$ to $A1$, where without an external force the profile simply changes back to $B1$ by releasing the same proton. But for a large force pointing to the left the shaft can rotate to the left by one c-subunit, and lose a proton through the port connected to the P-side. In this way protons can be transported from the N-side of the membrane to the P-side, at the expense of ATP hydrolysis by the F_1 component.

The transitions between the FEPs (at the regions where they overlap), i.e., the binding and release of protons, should even be faster than the transitions in the F_1 component, because the translocation of 3 or 4 protons is necessary for the synthesis of one ATP. How such large proton binding rates can be achieved at pH values around 7 is a big question [101]. Protonation is probably one of the rate-limiting processes of the enzyme's operation.

With these mechanochemical models we have given a *unified, quantitative description* for the operation of the ATP synthase, since the models for the two components can be readily connected by assuming that their common shaft is moving under the simultaneous action of the two FEPs. In an extended description this rigid coupling could be easily relaxed by introducing an elastic element between the two components. Though the models presented here are simplified and illustrative, they account for all the results of the mechanical experiments available at present. However, the final goal would be the determination of the entire FEPs in each possible chemical state of both components of the enzyme, together with the transition rate constants. This would be necessary and sufficient to describe the behaviour of the enzyme in all circumstances, indicating that this simple and transparent mechanochemical description gives the most suitable framework for studying the operation of the ATP synthase.

5.4 Collective effects

Recently, special attention has been paid to *collective effects* [105] arising from the interaction between a large number of Brownian motors that move along the same structure simultaneously. These studies have been motivated by various biological and physical systems: in muscle tissues hundreds of myosin molecules are attached to each other with their tails and form rigid myosin filaments that slide along the actin filaments. Similarly, in eukaryotic cilia and flagella rigidly attached dynein molecules drive the sliding of the microtubules. Experimental evidence shows that large groups of motor proteins can carry the same cargo and many free motors can move along the same microtubule. In motility assay studies a glass surface is coated with motor molecules that drive the track filaments. In separation processes a large number of particles move in the same medium.

The studies of ratchet systems with many interacting motors have revealed that collective effects lead to a wide range of new and complex features of the motion compared to the behaviour of a single particle. In all cases it has turned out that *commensurability* plays an essential role in the collective effects.

The motion of finite-sized Brownian particles interacting via hardcore repulsion in an asymmetric periodic potential have been investigated in both rocking [106] and flashing [107] ratchets. These two studies are presented in detail in §5.4.1 and §5.4.2, respectively. The main results are (i) the reversal of the direction of the motion as the particle density is increased, and (ii) a very strong and complex dependence of the average velocity on the size of the particles if the density is close to the maximum.

Jülicher and Prost [108] showed that the cooperation of a collection of motors, rigidly attached to each other and that independently adsorb to and desorb from a periodic structure, can lead to dynamical phase transitions and instabilities. In addition, the motors can generate a directed motion even if the system is symmetric, and the direction is selected by spontaneous symmetry breaking.

Csahók *et al.* [109] investigated the motion of a chain of elastically coupled particles in an asymmetric potential. The possibility of varying the strength of the interaction allowed them to study the crossover from single to collective motion.

5.4.1 Finite-sized particles in a 'rocking ratchet'

We present a simple one-dimensional model *via many interacting Brownian particles* moving with overdamped dynamics in a periodic potential [106]. According to the

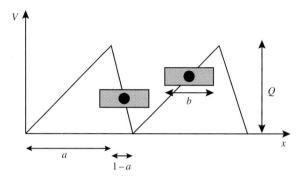

Fig. 5.27 Schematic picture of the system that we consider showing two particles with size b subject to the sawtooth-shaped periodic potential $V(x)$. The period of the potential is $\lambda = 1$, where the lengths of the slopes are $\lambda_1 = a$ and $\lambda_2 = 1 - a$. The potential difference between the top and the bottom is Q (after [106]).

computer simulations the model displays a number of novel cooperative phenomena: (i) first, as we show, for a range of frequencies of a periodic external driving force the average velocity v of the particles *changes its direction* as the number density of the particles is increased, and (ii) v has a sensitive dependence on the *size* of the migrating particles. This effect is demonstrated for two kinds of (periodic and constant) driving forces. We also present analytical results indicating that, in the case of a constant driving force, as the distance between the particles goes to zero the dependence of the velocity on the particle size becomes *extremely complex* (nondifferentiable).

The motion of a single particle (in the absence of other particles) is described by the Langevin equations (5.2):

$$\dot{x}_j = -\partial_x V(x_j) + F_j(t) + \sqrt{2k_B T} \xi_j(t), \quad j = 1, \ldots, N, \qquad (5.21)$$

where N is the number of particles, x_j denotes the position of the centre of mass of the jth particle, $V(x)$ is a sawtooth-shaped periodic potential, $F_j(t)$ is the fluctuating driving force with zero time average, and $\xi_j(t)$ is Gaussian white noise with the autocorrelation function $\langle \xi_j(t)\xi_i(t') \rangle = \delta_{j,i}\delta(t - t')$. The units are chosen so that the friction coefficient of the particles is 1. Since in most experimental situations the interaction between two particles can be well approximated by a hardcore repulsion, we assume that the *particles are hard rods* (see Fig. 5.27). The hardcore interaction means that during the motion the particles are not allowed to overlap: $(x_j - x_{j-1}) > b$ (i.e., a particle does not continue to move in its original direction if it touches another one). This rule complements eqn (5.21). All particles have the same size b, while the period of the potential is $\lambda = 1$. The size of the system or in other words the number of the periods is L. In the computer simulations we have applied periodic boundary conditions, and solved the finite difference version of eqn (5.21) numerically. $N \approx 20$ was usually large enough to get a good approximation for the current in the thermodynamic limit ($L, N \to \infty$ with L/N fixed).

The model is one-dimensional because the macromolecules serving as highways in biological transport can be assumed to be linear, representing well-defined tracks.

Thus, due to the hardcore interaction we also exclude the possibility of 'passing'. In higher dimensions (where the particles can get around each other, e.g., in separation techniques) further effects are expected to take place. In addition to the case of a periodic driving force we shall also consider the case of a constant driving force, because (i) the latter case is conceptually simpler, thus allowing a more direct interpretation of the simulation results and the analytic treatment of some limiting cases, and (ii) a zero-mean signal can always be constructed as an alternating ($+F$ and $-F$) piecewise-constant signal.

Normally, one single particle moves in the direction corresponding to the smaller uphill slope of the potential. However, there is a range of the parameters of the periodic driving force for which the particle migrates in the opposite direction [54–58]. In this regime we have found that gradual addition of particles in the system results in the change of their average velocity back to the 'normal' direction. We have tested this result for several different cases (including driving forces periodic in time [56] and distributed according to 'kangaroo' statistics [55]) and we have found that the change of the currents direction is a universal property of the collective motion in this model. Figure 5.28 shows a simple example, where the driving forces are $F_j(t) = A \sin(\omega_j t)$ and the ω_j values are chosen randomly around a fixed value ω with a dispersion of several percentage of ω (to avoid synchronization). The plot shows the average velocity as a function of ω, for various values of the particle density defined as $\rho = bN/L$ ($0 < \rho < 1$). In the inset we have plotted the fundamental diagram: the particle current $J \equiv vN/L$ as a function of the particle density for $\omega = 175$.

Another interesting feature is observed when the average separation distance between two neighbouring particles is fixed ($s \equiv L/N - b = $ const) and we change the size of the particles.

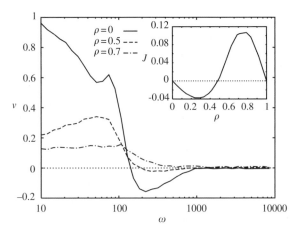

Fig. 5.28 The plot of the average velocity v as a function of the average frequency ω of the sinusoidal driving forces for three different values of the particle density $\rho \equiv bN/L$. The inset demonstrates the reversal of the particle current $J \equiv vN/L$ as a function of the particle density ρ, for $\omega = 175$. ($Q = 4$, $a = 0.8$, $b = 0.5$, $k_B T = 1$ and the amplitude of the driving forces $A = 32$) (after [106]).

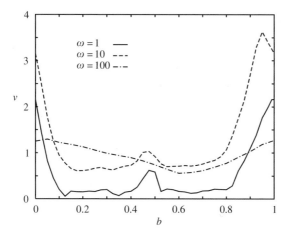

Fig. 5.29 The plot of the average velocity v as a function of the size of the particles b for three different values of the frequency ω of the sinusoidal driving forces. The average distance between two neighbouring particles is $s = 0.5$. ($Q = 4$, $a = 0.8$, $k_B T = 1$ and $A = 32$) (after [106]).

Before describing the results we mention that it is easy to show that a system of length L consisting of N particles of size $k+b$ ($0 \leq b < 1, k = 1, 2, \ldots$) is equivalent to a system of length $L - kN$ consisting of N particles of size b. Obviously, this kind of transformation has no effect on the motion of particles; therefore, it is enough to consider particles with sizes less than one. In other words, any quantity is a periodic function of the size of the particles with period 1, i.e., with period equal to the period of the underlying potential.

Figure 5.29 shows the average velocity as a function of the size of the particles in the above-mentioned case with sinusoidal driving forces for various values of ω. The velocity has very drastic changes. A large peak can be observed for b somewhat smaller than 1, and a smaller peak for b somewhat smaller than 1/2. In most of the other cases we have studied, a large peak is observed just before b reaches 1, or for b slightly larger than 0 (or equivalently slightly larger than 1), and a minimum (valley) on the opposite side of this integer value. This structure is repeated around 1/2, but on a smaller scale. Sometimes this structure can be observed around 1/3 and 2/3.

Investigating the origin of this strange behaviour of the particle size dependence on the average velocity we examine the simplest case when the driving force is stationary: $F_j(t) = F$ and smaller than the uphill gradient of the potential.

Let us consider the case when the size of the particles is somewhat less than 1 and there are two particles in neighbouring wells of the potential. Then the second particle is not able to jump further ahead until the first one jumps away. So the first one hinders the second one. Thus, the average velocity is smaller than the velocity of a single particle. Figure 5.30(a) shows this situation for 15 particles. A *vacancy-type* current can be observed, as a consequence of the traffic jams arising from the hindering of particles. This phenomenon is also related to jams common in one-dimensional driven diffusive systems and traffic models [110]. If the size of the particles is slightly larger

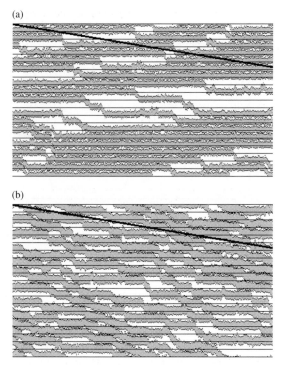

Fig. 5.30 Motion of the particles in the space–time domain. The time increases from left to right and the particles move downwards under the influence of the stationary driving force F. Horizontal lines represent the bottoms of the potential wells, and the wide slanted line represents the average motion of a single noninteracting particle. (a) Shows 15 particles with size $b = 0.833$. A vacancy-type current can be observed, as a consequence of the hindering effect of particles. The average velocity v is smaller than the average velocity of one single particle. (b) Shows 12 particles with size $b = 1.166$. There are no jams, and the density waves show that the particles assist each other in jumping over to the next well. v is larger than the velocity of a single particle. The average distance between particles is $s = 0.5$ in both cases (after [106]).

than 1 and there are also two particles in neighbouring wells, both of them cannot be at the minimum at the same time; therefore, the first one has a greater chance to jump further. In this case the second one indirectly 'pushes' the first one. (But the first one also hinders the second one.) Thus, in spite of the hindering effects the average velocity can be larger than the velocity of a single particle. This situation can be seen in Fig. 5.30(b) for 12 particles. There are no jams and the density waves show that the particles help each other to jump. In case of slowly alternating external forces these effects (hindering and pushing) are expected to influence the net transport.

Figure 5.31(a) shows the average velocity as a function of the size of the particles in this stationary case, for various values of the average distance $s \equiv L/N - b$ between two neighbouring particles. When s is infinity, the velocity is independent of the size of the particles and identical to the velocity of a single particle. On decreasing s a velocity peak starts to develop for s just larger than $b = 0$, and a valley appears for b close to, but smaller than, 1. This has been explained in the previous paragraphs.

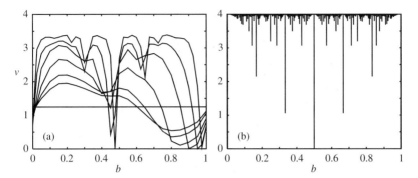

Fig. 5.31 The average velocity v as a function of the size of the particles b, when the driving force is stationary with $F = 4$. (a) The plot for different values of the average distance between two neighbouring particles: $s = \infty$ (one single particle, the horizontal line) and $s = 0.6, 0.4, 0.2, 0.1, 0.05, 0.025$. (b) The plot in the limit when the average distance between two particles goes to zero. This discontinuous function has sharp minima if b is rational and a value equal to F if b is irrational. ($Q = 4$, $a = 0.2$ and $k_B T = 1$) (after [106].)

As s is further decreased, another peak appears beyond $b = 1/2$ and also a valley before $b = 1/2$. This can also be explained in the above manner, taking into consideration that two particles can sit in the same potential well if $b \approx 1/2$, and we can handle them as one particle with size $b \approx 1$. Decreasing the average distance further, valleys and peaks appear before and after $b = 1/3$ (3 particles in 1 potential well), $b = 2/3$ (3 particles in 2 potential wells) and so on at (almost) any rational value of b.

If the sum of the average distance between the neighbouring particles and the size of the particles is a rational value, i.e., $b + s = n/m$ (where n and m are integers), we can say that the structure is *commensurate*. For $s \ll 1$ the particles are distributed evenly and m particles can be found in n potential wells. The minimum of the potential energy of the system is realized if every mth particle is sitting in the bottom of the potential wells. Then, for $F = 0$ each particle has to jump a distance $1/m$ to reach the next minimum energy state of the system. Simple algebra shows that such a system (in which N particles play the role of a single particle) can also be described in terms of a modified (rescaled) sawtooth-shaped potential, where the lengths of the slopes are $\lambda_1' = \{ma\}/m$ and $\lambda_2' = \{m(1-a)\}/m$, and the period is $\lambda' = \lambda_1' + \lambda_2' = 1/m$. The potential difference between the top and the bottom of the wells of the rescaled potential is NQ', where

$$Q' = Q \frac{\{ma\}}{ma} \frac{\{m(1-a)\}}{m(1-a)}. \tag{5.22}$$

The notation $\{\ldots\}$ means the fractional part of the value between the braces.

Thus, in the presence of a driving force F we can calculate the average velocity of the particles as the velocity of a single particle in the rescaled potential using the formula derived by Magnasco [53]:

$$v(F) = \frac{\lambda P_2^2 \sinh\left(\frac{F\lambda}{2k_B T}\right)}{k_B T \left(\frac{\lambda}{Q}\right)^2 \left[\cosh\left(\frac{Q - F\Delta/2}{k_B T}\right) - \cosh\left(\frac{F\lambda}{2k_B T}\right)\right] - P_1 P_2 \left(\frac{\lambda}{Q}\right)\sinh\left(\frac{F\lambda}{2k_B T}\right)},$$
(5.23)

$$P_1 = \Delta + \frac{\lambda^2 - \Delta^2}{4}\frac{F}{Q}, \qquad P_2 = \left(1 - \frac{F\Delta}{2Q}\right)^2 - \left(\frac{F\Delta}{2Q}\right)^2,$$
(5.24)

$$\lambda = \lambda_1 + \lambda_2, \qquad \Delta = \lambda_1 - \lambda_2.$$
(5.25)

With the rescaled values of the parameters: $\{Q, \lambda_1, \lambda_2, F, T\} \to \{Q', \lambda_1', \lambda_2', F, T/N\}$. Note that the rescaled temperature goes to zero as N goes to infinity.

If, on the other hand, the structure is incommensurate and the average distance between the particles is small, the rescaled potential becomes flat and the system has a continuous translation symmetry. Thus, the particles can move with the maximum velocity $v_{max} = F$.

The modified potential is also flat if the structure is commensurate but ma is an integer number. This is the reason why we cannot see valleys near 1/5, 2/5, 3/5 and 4/5 in Fig. 5.31(a) for $a = 0.2$.

Correspondingly, on decreasing the average distance between the particles the minima of the valleys converge to the rational values of the b axis, the values at the minima converge to the values calculated from Magnasco's formula and the width of the valleys goes to zero. For all other values of b the velocity goes to $v_{max} = F$. In the limit when the average distance goes to zero, we get a strange, discontinuous function with sharp minima at the rational values of b but a constant F elsewhere (Fig. 5.31(b)).

5.4.2 Finite-sized particles in a 'flashing ratchet'

Now we study the motion of particles with hardcore repulsion, in an asymmetric potential that is switched on and off *simultaneously* for all the particles [107]. This natural extension to many particles of the on/off model of Ajdari and Prost [49] also leads to a rich phenomenology. In particular, for long enough duration of the 'off' intervals, we prove analytically that the average velocity can change sign a few times as the particle density is increased from 0 to 1, and that its sign and amplitude at high density is extremely sensitive to the particles' size.

We consider the overdamped motion of N Brownian particles of size $0 \leq b < 1$ on a segment of length L, as previously. They are submitted to a sawtooth periodic potential $V(x, t)$, which is periodically turned 'on' ($V = V_{on}(x)$) for a time τ_{on} and then 'off' for a time τ_{off} ($V = V_{off} = 0$) (see Fig. 5.32). The asymmetry of the potential is characterized by the length a of its steeper slope. The motion of the particles is governed by the Langevin equations (5.21) with the extension that the potential V is time dependent.

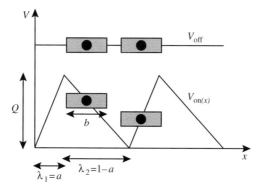

Fig. 5.32 Schematic picture of the system showing two particles of size b submitted periodically to the sawtooth periodic potential $V_{on}(x)$ for a time τ_{on}, and then to the flat potential V_{off} for a time τ_{off}. The sawtooth potential has a period $\lambda = 1$, the sum of a short size $\lambda_1 = a$ and a long one $\lambda_2 = 1 - a$. The corresponding energy barrier is Q (after [107]).

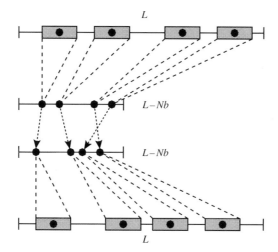

Fig. 5.33 During the motion of the particles in the flat potential V_{off}, the real system can be reduced to a system of size $L - Nb$ with N noninteracting point-like particles. The intervals between the particles are kept constant when switching from one system to the other (after [107]).

Before going further let us briefly comment on a simple way to describe the diffusive 'off' stage in the general case (see Fig. 5.33). The system can be 'compressed' to a set of N *point-like* particles in a system of size $L - Nb$, keeping the intervals between particles unchanged. When two point-like particles meet we can either let them cross and swap (rename) them afterwards, or forbid crossing; the choice does not affect their final position. This means that the particles can be handled as *noninteracting* ones as long as they are reordered at the end of the 'off' stage. Inverting the 'compression' procedure leads back to the original system.

Note that the 'compression' picture of point-like particles is valid in the continuous limit, when the underlying microscopic jump distance δ_{mic} and time t_{mic} go to zero with the diffusion coefficient δ_{mic}^2/t_{mic} kept fixed. For very small but finite δ_{mic} and t_{mic} this picture fails to describe the freezing of very large aggregates of 'touching' particles. Our description thus holds up to a 'cap' density (that tends towards 1 as δ_{mic} and t_{mic} go to zero), beyond which the effective velocity will decrease to zero.

Let us now investigate how the average velocity v of the particles depends on their size b and density $\rho \equiv bN/L$, in the limit of large systems (N and L go to infinity while ρ remains finite). To get analytical solutions we focus on specific regimes, as was done for the simpler, one particle limit $\rho \to 0$ [49]. First, the pinning potential V_{on} is taken to be strong enough so that during the time τ_{on} the particles drift quickly to the positions corresponding to the nearest local energy minimum of the system, where they get trapped. This deep potential-well limit ($Q \gg k_B T$) furthermore suits fast separation purposes. Second, most of the results will be obtained in the limit where τ_{off} is long enough for the particles to forget (modulo the period) their initial position on the sawtooth during an 'off' period. The average displacement over a cycle is then that of initially randomly distributed particles during a single 'on' phase.

In the low-density limit, a particle with random initial position in the $[-\lambda_2, \lambda_1]$ period ends at $x = 0$ after an 'off' phase. The average progression per cycle is thus $\langle d \rangle = \frac{1}{2}(\lambda_2 - \lambda_1) = \frac{1}{2} - a$. Let us now turn to the other extreme: an almost-packed system $\rho \approx 1$. As in §5.4.1, the *commensuration* of the particle system to the potential period plays a crucial role.

To see this, consider the limit case $\rho = 1$ ($L/N = b$), where the system is equivalent to a single particle of size L, the position of which is measured by x_1 for example. Take now the limit of a very large system: $N \to \infty$. In the incommensurate case (b irrational), the particles are then uniformly distributed in the periods whatever the value of x_1, so that the whole system feels a flat potential whether the sawtooth potential is 'on' or 'off': the average velocity of the particles is zero. This is to be contrasted to the case of Jülicher and Prost [108] (see §5.4.3) where, as particles switched between 'on' and 'off' independently, motion could be obtained in the incommensurate situation. We now turn to the much richer commensurate case: $b = n/m$ in irreducible form. As in §5.4.1, the effective potential seen by the equivalent L-size particle during 'on' periods is a sawtooth potential of period $\lambda' = 1/m$ with two linear pieces of lengths $\lambda'_1 = \{ma\}/m$ and $\lambda'_2 = \{m(1 - a)\}/m$, and the barrier height NQ' is given by eqn (5.22). Applying the single particle limit to the equivalent particle, we get its average displacement per cycle (which is that of every real particle):

$$\langle d \rangle = \frac{1}{2}(\lambda'_2 - \lambda'_1) = \frac{1}{2m}(1 - 2\{ma\}). \tag{5.26}$$

A formal problem with the above analysis is that the diffusion coefficient of the equivalent particle is $k_B T/N$ so that randomization in the 'off' phase cannot be achieved in a finite time τ_{off} (in the limit $N \to \infty$, $\rho = 1$). Consider instead the commensurate case with a density $\rho = 1 - \varepsilon$, $0 < \varepsilon \ll 1$, and take first the limit $N \to \infty$. What are the differences with the $\rho = 1$ case? At the end of an 'on' pinning stage, there are now a few very distant empty spaces at the top of some potential barriers, the sizes of which are usually not greater than b, that separate groups of $\simeq 1/\varepsilon$ 'touching' particles. This gives the initial conditions for the following diffusing stage: in the 'compressed' picture (Fig. 5.33), each group consists of many point-like particles located at the same position, separated by distances of order b. So if the 'off' time allows a free particle to diffuse over distances of order b, the particles will be randomly distributed (typical separation $\simeq \varepsilon b$). Upon switching the sawtooth potential 'on', to zeroth order in ε the average displacement is $\langle d \rangle$ as given by eqn (5.26). Thus, the average velocity tends towards $\simeq \langle d \rangle/(\tau_{\text{on}} + \tau_{\text{off}})$ as ε goes to zero, τ_{off} being kept constant but large enough to allow free diffusion over distances larger than b: $\tau_{\text{off}} > b^2/(k_B T)$ in our units. A crucial point of this argument is that N should be larger than $1/\varepsilon$, indicating that $N \to \infty$, $\rho \to 1$ is a singular limit

This leads to a quite strange behaviour for the high-density ($N \to \infty$, $\rho = 1 - \varepsilon$) drift as a function of the particle size, as illustrated by Fig. 5.34 (similarly as in Fig. 5.31). The limit average displacement per on-off cycle $\langle d \rangle$ is an erratic, discontinuous function with sharp peaks (given by eqn (5.26)) for rational values of b, and

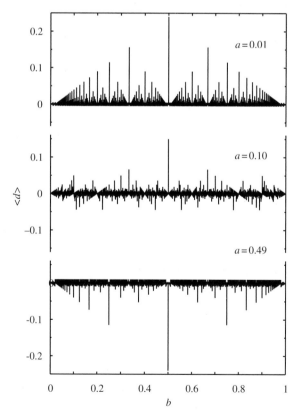

Fig. 5.34 High-density limit for the average particle displacement per on-off cycle $\langle d \rangle$ as a function of the particle size b, for different values of the asymmetry parameter a (after [107]).

zero otherwise. Both positive and negative peaks are present, in a pattern that depends on the value of the asymmetry parameter a.

Remember that at zero particle density, the average displacement per cycle is positive (if $a < 0.5$), and independent of the size of the particles. Increasing the density from 0 to 1, causes it to evolve to the discontinuous function of b shown in Fig. 5.34. The question is how this occurs. When the density is still less than one, the function $\langle d(b) \rangle$ has to be continuous, due to the smoothing effect of the finite temperature. But we will now show that by increasing the density from 0 to 1 while keeping the particle size b fixed, the velocity can vary nonmonotonically and can even change sign several times, along a route that will be sensitive to the actual value of b.

Let us start with a simple pedagogical example: $b = \frac{1}{3}$ and $a = 0.5 - b/4$. For very small particle densities, a particle is usually alone in some period of the potential. Its average displacement per on–off cycle is about $b/4$, as it is the distance between its eventual position (the bottom of the well) and its average starting position (the middle of the well). At larger densities, when on average two particles fall into a well, they will end in the configuration of minimal potential energy: the two particles touch each other, with the centre of the right particle in the bottom of the well. Thus, the centre of mass of the two particles is $b/4$ far from the middle of the well to the left. Therefore $\langle d \rangle$, the average drift during the 'on' stage, is now about $-b/4$, a 'negative' value! At even higher densities, with about three particles per well, the average displacement will be about $b/4$ again.

Analytical calculations are actually possible for $b = 1/m$ with $m = 2, 3, \ldots,$ as the wells are then 'independent': when the potential is switched 'on', the particles remain in their starting well, and the minimal energy configuration in each well is independent of what happens in the neighbouring ones. Always assuming that the particles are randomized before each 'on' period, one can calculate the probabilities that $0, 1, \ldots, m$ particles will start in the same well (the intervals between particles follow a Poisson distribution). By determining the corresponding minimal energy configurations at the end of the 'on' stage, $\langle d \rangle$ can then be computed explicitly. This straightforward procedure becomes tedious for large m and leads to long formulas. Curves are plotted in Fig. 5.35, and we only give the simplest formula:

$$\langle d \rangle = \frac{1}{4} + \frac{1}{4}\left(\frac{1}{\rho} - 1\right)\left(1 - \exp\left(-\frac{\rho}{1 - \rho}\right)\right) - a, \qquad (5.27)$$

for $b = 1/2$ and $0 \leq a < 0.5$. It is actually easy to see that for any rational value of b ($b = n/m$), the quantity $\langle d(\rho, a) \rangle + a$ does not depend on a within intervals where the minimal energy configurations are the same. Therefore, in Fig. 5.35, we have plotted $\langle d \rangle + a$ as a function of ρ.

The figure shows that the average displacement (or velocity) of the particles is a nonmonotonous function of the density. For well-chosen values of a (e.g., $a \approx 0.5 - b/4$) the direction of motion can even change several times. The curves obtained are consistent with both the low-density result $\langle d \rangle + a = \frac{1}{2}$ and with our high-density ($\rho \to 1$) estimate as quantified by eqn (5.26).

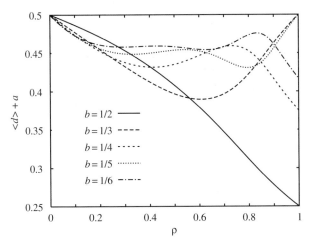

Fig. 5.35 The plot of $\langle d \rangle + a$ as a function of the density ρ for five different values of b, valid if a lies in the appropriate interval, which is respectively [0,1/2], [1/3,1/2], [1/3,1/2], [2/5,1/2], and [2/5,1/2] in decreasing order of b (after [107]).

Eventually, to investigate the range of validity of the analytical results, we performed numerical simulations (using the compression picture), and present here (Fig. 5.36) the case $b = \frac{1}{5}$, $a = 0.5 - b/4 = 0.45$, $k_B T = 1$, for different values of τ_{off}. For large values of τ_{off} (e.g., $\tau_{\text{off}} = 0.3$), the simulation result is indistinguishable from the analytic predictions. Analysis also shows that the time for the randomization in the 'off' state is larger at low density, in agreement with our estimate that it should scale as λ^2 at low density but as b^2 at high density. More importantly, Fig. 5.36 shows that the high-density velocities $v = \langle d \rangle / (\tau_{\text{on}} + \tau_{\text{off}})$ can be larger than the low-density ones: particles in the 'off' state need not diffuse over a distance a but only over b, which allows us to reduce the cycle time.

5.4.3 Collective behaviour of rigidly attached particles

In several biological studies large groups of motor proteins work together to transport relatively big objects. Jülicher and Prost [108] introduced a model to describe the behaviour of these systems. In this model large number of particles are subject to a periodic, asymmetric potential that is turned 'on' and 'off' independently for each particle, and the particles are attached to a rigid backbone that keeps the distance between them fixed (Fig. 5.37).

In the following, we focus on the case where either the particles are randomly attached to the backbone or the spacing between them is constant but incommensurate with the period of the potential. In both cases the particles are uniformly distributed in the periods, and can be characterized by the probabilities $P_{\text{on}}(x, t)$ and $P_{\text{off}}(x, t) = 1 - P_{\text{on}}(x, t)$ showing which fractions of the particles are in the 'on' and 'off' states

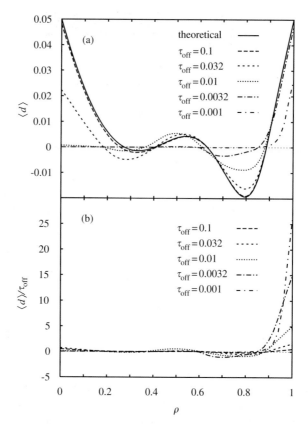

Fig. 5.36 The plots (a) and (b) show respectively $\langle d \rangle$ and $\langle d \rangle / \tau_{\text{off}}$ as a function of the density ρ in the case $b = 1/5$, $a = 0.5 - b/4 = 0.45$, $k_B T = 1$, for different values of τ_{off}. Supposing that Q is large, τ_{on} can be chosen small enough ($\tau_{\text{on}} \ll \tau_{\text{off}}$) so that $\langle d \rangle / \tau_{\text{off}}$ is a good approximation to the velocity $v = \langle d \rangle / (\tau_{\text{on}} + \tau_{\text{off}})$ (after [107]).

of the potential respectively at time t and position x ($0 \leq x < 1$). The equations of motion for this system are [108]

$$\partial_t P_{\text{on}} + v \partial_x P_{\text{on}} = -\omega_{\text{on}}(x) P_{\text{on}} + \omega_{\text{off}}(x) P_{\text{off}},$$
$$\partial_t P_{\text{off}} + v \partial_x P_{\text{off}} = +\omega_{\text{on}}(x) P_{\text{on}} - \omega_{\text{off}}(x) P_{\text{off}}, \tag{5.28}$$

where $\omega_{\text{on}}(x)$ and $\omega_{\text{off}}(x)$ denote the transition rate constants between the two states of the potential, and the velocity of the backbone, v, is determined by $v = f_{\text{ext}} + f$ (the units are chosen again so that the friction coefficient is 1). The external force f_{ext} and the average force

$$f = -\int_0^1 dx \, (P_{\text{on}} \partial_x V_{\text{on}}(x) + P_{\text{off}} \partial_x V_{\text{off}}(x)) \tag{5.29}$$

exerted by the potentials (which should be zero for the flat potential) are normalized per particle.

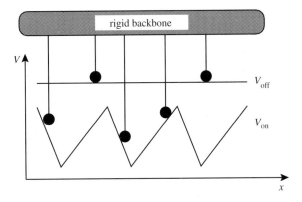

Fig. 5.37 Schematic representation of the two-state model with many particles attached to a rigid backbone (after [105]).

In the steady state, using the relation $P_{\text{off}} = 1 - P_{\text{on}}$, these equations reduce to

$$v\partial_x P_{\text{on}} = -(\omega_{\text{on}}(x) + \omega_{\text{off}}(x))P_{\text{on}} + \omega_{\text{off}}(x), \tag{5.30}$$

$$f_{\text{ext}} = v + \int_0^1 dx\, P_{\text{on}}\partial_x (V_{\text{on}}(x) - V_{\text{off}}(x)), \tag{5.31}$$

allowing the determination of the external force $f_{\text{ext}}(v)$ as a function of the velocity v of the backbone. Equation (5.30) can be solved either analytically for some potential shapes or in a power expansion as a function of v [108].

Let us express $\omega_{\text{on}}(x)$ as $\omega_{\text{off}}(x)e^{(V_{\text{on}}(x)-V_{\text{off}}(x))/(k_B T)} + \Omega\Theta(x)$, where now the 'amplitude' Ω measures the distance from equilibrium (for $\Omega = 0$ detailed balance holds). In order to discuss collective effects, we consider a system with no external force and with symmetric periodic potentials. Let the perturbation $\Theta(x)$ be nonnegative, also symmetric, and localized in the vicinity of the potential minimum (i.e., it differs from zero only in that region). It can be shown [108] that if Ω is smaller than a critical value Ω_c, the velocity of the backbone is zero, but for $\Omega > \Omega_c$ the solution bifurcates into two stable solutions $v = \pm v(\Omega)$, while the $v = 0$ solution becomes unstable. Therefore, a *continuous onset of motion* occurs at $\Omega = \Omega_c$ via *spontaneous symmetry breaking*.

This spontaneous symmetry breaking can be understood qualitatively as follows. For $v = 0$ the localized excitation $\Omega\Theta(x)$ leads to a depletion of $P_{\text{on}}(x)$ near the potential minimum. Since $\Theta(x)$ and the potentials are spatially symmetric, the force f vanishes. If the system is now perturbed in such a way that the backbone moves to the right with a small velocity v, the depletion of $P_{\text{on}}(x)$ is transported to the right. Now the population along the positive potential slope is depleted, while along the negative slope it has gained particles. As a result, the average force f pulls the backbone to the right and increases the effect of the initial perturbation.

Another version of this model, where the backbone is elastically coupled to a fixed object, leads to the onset of spontaneous oscillations instead of unidirectional motion [111].

References

[1] J. Darnell, H. Lodish, and D. Baltimore. *Molecular Cell Biology*. New York, Scientific American Books, 1990.

[2] B. Alberts, D. Bray, M. Raff, K. Roberts, and J. Watson. *Molecular Biology of the Cell*. New York, Garland, 1994.

[3] B. G. Levi. Measured steps advance the understanding of molecular motors. *Phys. Today* 48: 17–19, 1995.

[4] Y.-H. Song and E. Mandelkow. Recombinant kinesin motor domain binds to beta-tubulin and decorates microtubules with a B surface lattice. *Proc. Natl. Acad. Sci. USA* 90: 1671–1675, 1993.

[5] A. Hoenger, E. P. Sablin, R. D. Vale, R. J. Fletterick, and R. A. Milligan. Three-dimensional structure of a tubulin–motor-protein complex. *Nature* 376: 271–274, 1995.

[6] M. Kikkawa, T. Ishikawa, T. Wakabayashi, and N. Hirokawa. Three-dimensional structure of the kinesin head–microtubule complex. *Nature* 376: 274–277, 1995.

[7] K. Hirose, A. Lockhart, R. A. Cross, and L. A. Amos. Nucleotide-dependent angular change in kinesin motor domain bound to tubulin. *Nature* 376: 277–279, 1995.

[8] A. Ashkin, K. Schütze, J. M. Dziedzic, U. Euteneuer, and M. Schliwa. Force generation of organelle transport measured in vivo by an infrared laser trap. *Nature* 348: 346–348, 1990.

[9] R. D. Vale, T. S. Reese, and M. P. Sheetz. Identification of a novel force-generating protein, kinesin, involved in microtubule-based motility. *Cell* 42: 39–50, 1985.

[10] F. J. Kull, E. P. Sablin, R. Lau, R. J. Fletterick, and R. D. Vale. Crystal structure of the kinesin motor domain reveals a structural similarity to myosin. *Nature* 380: 550–555, 1996.

[11] E. P. Sablin, F. J. Kull, R. Cooke, R. D. Vale, and R. J. Fletterick. Crystal structure of the motor domain of the kinesin-related motor NCD. *Nature* 380: 555–559, 1996.

[12] S. Ray, E. Meyhöfer, R. A. Milligan, and J. Howard. Kinesin follows the microtubule's protofilament axis. *J. Cell. Biol.* 121: 1083–1093, 1993.

[13] A. J. Hunt, F. Gittes, and J. Howard. The force exerted by a single kinesin molecule against a viscous load. *Biophys. J.* 67: 766–781, 1994.

[14] K. Svoboda and S. M. Block. Force and velocity measured for single kinesin molecules. *Cell* 77: 773–784, 1994.

[15] K. Svoboda, C. F. Schmidt, B. J. Schnapp, and S. M. Block. Direct observation of kinesin stepping by optical trapping interferometry. *Nature* 365: 721–727, 1993.

[16] C. M. Coppin, D. W. Pierce, L. Hsu, and R. D. Vale. The load dependence of kinesin's mechanical cycle. *Proc. Natl. Acad. Sci. USA* 94: 8529–8544, 1997.

[17] M. J. Schnitzer and S. M. Block. Kinesin hydrolyses one ATP per 8 nm step. *Nature* 388: 386–390, 1997.

[18] K. Svoboda, P. P. Mitra, and S. M. Block. Fluctuation analysis of motor protein movement and single enzyme kinetics. *Proc. Natl. Acad. Sci. USA* 91: 11782–11786, 1994.

[19] K. Visscher, M. J. Schnitzer, and S. M. Block. Single kinesin molecules studied with a molecular force clamp. *Nature* 400: 184–189, 1999.

[20] U. Henningsen and M. Schliwa. Reversal in the direction of movement of a molecular motor. *Nature* 389: 93–96, 1997.

[21] S. A. Endow and K. W. Waligora. Determinants of kinesin motor polarity. *Science* 281: 1200–1202, 1998.

[22] E. P. Sablin, R. B. Case, S. C. Dai, C. L. Hart, A. Ruby, R. D. Vale, and R. J. Fletterick. Direction determination in the minus-end-directed kinesin motor ncd. *Nature* 395: 813–816, 1998.

[23] W. Bloom and D. W. Fawcett, *A Textbook of Histology (2nd edn)*. Philadelphia, W.B. Saunders, 1975.

[24] R. W. Lymn and E. W. Taylor. Mechanism of adenosine triphosphate hydrolysis by actomyosin. *Biochemistry* 10: 4617–4624, 1971.

[25] Y. E. Goldman. Kinetics of actomyosin ATPase in muscle fibers. *Ann. Rev. Physiol.* 49: 629–654, 1987.

[26] C. R. Bagshaw. *Muscle Contraction*. London, Chapman & Hall, 1993.

[27] J. T. Finer, R. M. Simmons, and J. A. Spudich. Single myosin molecule mechanics: piconewton forces and nanometre steps. *Nature* 368: 113–119, 1994.

[28] I. Rayment, W. R. Rypniewski, K. Schmidtbase, R. Smith, D. R. Tomchick, M. M. Benning, and D. A. Winkelmann. 3-dimensional structure of myosin subfragment-1 – a molecular motor. *Science* 261: 50–58, 1993.

[29] I. Rayment, H. M. Holden, M. Whittaker, C. B. Yohn, M. Lorenz, K. C. Holmes, and R. A. Milligan. Structure of the actin-myosin complex and its implication for muscle contraction. *Science* 261: 58–65, 1993.

[30] A. J. Fisher, C. A. Smith, J. Thoden, R. Smith, K. Sutoh, H. M. Holden, and I. Rayment. Structural studies of myosin-nucleotide complexes: a revised model for the molecular basis of muscle contraction. *Biophys. J.* 68: 19s–28s, 1995.

[31] R. Dominguez, Y. Freyzon, K. M. Trybus, and C. Cohen. Crystal structure of a vertebrate smooth muscle myosin motor domain and its complex with the essential light chain: visualization of the pre-power stroke state. *Cell* 94: 559–571, 1998.

[32] E. M. Ostap and D. D. Thomas. Transient detection of spin-labeled myosin's internal dynamics and global orientation during ATP hydrolysis. *Biophys. J.* 68: 335s, 1995.

[33] J. E. T. Corrie, B. D. Brandmeier, R. E. Ferguson, D. R. Trentham, J. Kendrick-Jones, S. C. Hopkins, U. A. van der Heide, Y. E. Goldman, C. Sabido-David,

R. E. Dale, S. Criddle, and M. Irving. Dynamic measurement of myosin light-chain-domain tilt and twist in muscle contraction. *Nature* 400: 425–430, 1999.

[34] P. D. Boyer. The binding change mechanism for ATP synthase – some probabilities and possibilities. *Biochim. Biophys. Acta* 1140: 215–250, 1993.

[35] R. H. Fillingame. Coupling H^+ transport and ATP synthesis in F_1F_0-ATP synthases: glimpses of interacting parts in a dynamic molecular machine. *J. Exp. Biol.* 200: 217–224, 1997.

[36] R. Cross and T. Duncan. Subunit rotation in F_1F_0-ATP synthases as a means of coupling proton transport through F_0 to the binding changes in F_1. *J. Bioenerg. Biomem.* 28: 403–408, 1996.

[37] W. Junge, H. Lill, and S. Engelbrecht. ATP synthase: An electrochemical transducer with rotatory mechanics. *Trends Biochem. Sci.* 22: 420–423, 1997.

[38] D. Sabbert, S. Engelbrecht, and W. Junge. Intersubunit rotation in active F-ATPase. *Nature* 381: 623–625, 1996.

[39] H. Noji, R. Yasuda, M. Yoshida, and K. Kinosita. Direct observation of the rotation of F_1-ATPase. *Nature* 386: 299–302, 1997.

[40] J. P. Abrahams, A. G. W. Leslie, R. Lutter, and J. Walker. Structure at 2.8 Å resolution of F_1-ATPase from bovine heart mitochondria. *Nature* 370: 621–628, 1994.

[41] S. J. Kron and J. A. Spudich. Fluorescent actin filaments move on myosin fixed to a glass surface. *Proc. Natl. Acad. Sci. USA* 83: 6272–6276, 1986.

[42] L. Bourdieu, M. O. Magnasco, D. A. Winkelmann, and A. Libchaber. Actin filaments on myosin beds: the velocity distribution. *Phys. Rev. E* 52: 6573–6579, 1995.

[43] R. D. Astumian and M. Bier. Mechanochemical coupling of the motion of molecular motors to ATP hydrolysis. *Biophys. J.* 70: 637–653, 1996.

[44] P. Hänggi and R. Bartussek. Brownian rectifiers: how to convert Brownian motion into directed transport. In *Nonlinear Physics of Complex Systems, Lecture Notes in Physics* (ed. J. Parisi, S. C. Müller, and W. Zimmermann). 467: 294–308. Springer, Berlin, 1996.

[45] R. D. Astumian. Thermodynamics and kinetics of a Brownian motor. *Science* 276: 917–922, 1997.

[46] F. Jülicher, A. Ajdari, and J. Prost. Modelling molecular motors. *Rev. Mod. Phys.* 69: 1269–1281, 1997.

[47] I. Derényi and T. Vicsek. Realistic models of biological motion. *Physica A* 249: 397–406, 1998.

[48] R. P. Feynman, R. B. Leighton, and M. Sands. *The Feynman Lectures on Physics*. Addison-Wesley, Reading, MA, 1966.

[49] A. Ajdari and J. Prost. Drift induced by a spatially periodic potential of low symmetry: pulsed dielectrophoresis. *C. R. Acad. Sci. Paris* 315: 1635–1639, 1992.

[50] R. D. Astumian and M. Bier. Fluctuation-driven ratchets: molecular motors. *Phys. Rev. Lett.* 72: 1766–1769, 1994.

[51] J. Prost, J.-F. Chauwin, L. Peliti, and A. Ajdari. Asymmetric pumping of particles. *Phys. Rev. Lett.* 72: 2652–2655, 1994.

[52] C. S. Peskin, G. B. Ermentrout, and G. F. Oster. The correlation ratchet: a novel mechanism for generating directed motion by ATP hydrolysis. In *Cell Mechanics and Cellular Engineering* (ed. V. C. Mow *et al.*). Springer-Verlag, New York, 1994.

[53] M. O. Magnasco. Forced thermal ratchets. *Phys. Rev. Lett.* 71: 1477–1480, 1993.

[54] M. M. Millonas and D. I. Dykman. Transport and current reversal in stochastically driven ratchets. *Phys. Lett. A* 185: 65–69, 1994.

[55] C. R. Doering, W. Horsthemke, and J. Riordan. Nonequilibrium fluctuation-induced transport. *Phys. Rev. Lett.* 72: 2984–2987, 1994.

[56] R. Bartussek, P. Hänggi, and J. G. Kissner. Periodically rocked thermal ratchets. *Europhys. Lett.* 28: 459–464, 1994.

[57] M. M. Millonas. Self-consistent microscopic theory of fluctuation-induced transport. *Phys. Rev. Lett.* 74: 10–13, 1995.

[58] R. Bartussek, P. Reimann, and P. Hänggi. Precise numerics versus theory for correlation ratchets. *Phys. Rev. Lett.* 76: 1166–1169, 1996.

[59] A. Ajdari, D. Mukamel, L. Peliti, and J. Prost. Rectified motion induced by ac forces in periodic structures. *J. Phys. I, France* 4: 1551–1561, 1994.

[60] D. R. Chialvo and M. M. Millonas. Asymmetric unbiased fluctuations are sufficient for the operation of a correlation ratchet. *Phys. Lett. A* 209: 26–30, 1995.

[61] J. Łuczka, R. Bartussek, and P. Hänggi. White noise-induced transport in spatially periodic structures. *Europhys. Lett.* 31: 431–436, 1995.

[62] J.-F. Chauwin, A. Ajdari, and J. Prost. Force-free motion in asymmetric structures: a mechanism without diffusive steps. *Europhys. Lett.* 27: 421–426, 1994.

[63] J.-F. Chauwin, A. Ajdari, and J. Prost. Current reversal in asymmetric pumping. *Europhys. Lett.* 32: 373–378, 1995.

[64] M. B. Tarlie and R. D. Astumian. Optimal modulation of a Brownian ratchet and enhanced sensitivity to a weak external force. *Proc. Natl. Acad. Sci. USA* 95: 2039–2043, 1998.

[65] R. Landauer. Motion out of noisy states. *J. Stat. Phys.* 53: 233–248, 1988.

[66] M. Büttiker. Transport as a consequence of state-dependent diffusion. *Z. Phys. B* 68: 161–167, 1987.

[67] I. Derényi and R. D. Astumian. Efficiency of Brownian heat engines. *Phys. Rev. E* 59: R6219–R6222, 1999.

[68] S. M. Block and K. Svoboda. Analysis of high-resolution recordings of motor movement. *Biophys. J.* 68: 230s–241s, 1995.

[69] I. Derényi and T. Vicsek. The kinesin walk: a dynamic model with elastically coupled heads. *Proc. Natl. Acad. Sci. USA* 93: 6775–6779, 1996.

[70] A. Ajdari. Force-free motion in an asymmetric environment: a simple model for structured objects. *J. Phys. I, France* 4: 1577–1582, 1994.

[71] S. P. Gilbert, M. R. Webb, M. Brune, and K. A. Johnson. Kinesin cross-bridge detachment occurs after ATP hydrolysis. *Nature* 373: 671–676, 1995.

[72] W. O. Fenn. The relation between the work performed and the energy liberated in muscular contraction. *J. Physiol.* 184: 373–395, 1924.

[73] C. S. Peskin and G. Oster. Coordinated hydrolysis explains the mechanical behavior of kinesin. *Biophys. J.* 68: 202s–211s, 1995.

[74] T. Duke and S. Leibler. Motor protein mechanics: A stochastic model with minimal mechanochemical coupling. *Biophys. J.* 71: 1235–1247, 1996.

[75] R. D. Astumian and I. Derényi. Fluctuation-driven transport and models of molecular motors and pumps. *Eur. Biophys. J.* 27: 474–489, 1998.

[76] R. D. Astumian and I. Derényi. A chemically reversible brownian motor: application to kinesin and ncd. *Biophys. J.* 77: 993–1002, 1999.

[77] Y. Okada and N. Hirokawa. A processive single-headed motor: kinesin superfamily protein KIF1A. *Science* 283: 1152–1157, 1999.

[78] A. F. Huxley. Muscle structure and theories of contraction. *Prog. Biophys.* 7: 255–318, 1957.

[79] T. L. Hill. Theoretical formalism for the sliding filament model of contraction of striated muscle, part I. *Prog. Biophys. Mol. Biol.* 28: 267–340, 1974.

[80] E. Eisenberg, T. L. Hill, and Y. Chen. Cross-bridge model of muscle contraction. *Biophys. J.* 29: 195–227, 1980.

[81] E. Pate and R. Cooke. A model of cross-bridge action: the effect of ATP, ADP and P_i. *J. Muscle Res. Cell Motil.* 10: 181–196, 1989.

[82] E. Pate, H. White, and R. Cooke. Determination of the myosin step size from mechanical and kinetic data. *Proc. Natl. Acad. Sci. USA* 90: 2451–2455, 1993.

[83] N. J. Córdova, B. Ermentrout, and G. F. Oster. Dynamics of single-motor molecules: the thermal ratchet model. *Proc. Natl. Acad. Sci. USA* 89: 339–343, 1992.

[84] S. Leibler and D. A. Huse. Porters versus rowers: a unified stochastic model of motor proteins. *J. Cell. Biol.* 121: 1357–1368, 1993.

[85] D. A. Smith and M. A. Geeves. Strain-dependent cross-bridge cycle for muscle. *Biophys. J.* 69: 524–537, 1995.

[86] D. A. Smith and M. A. Geeves. Strain-dependent cross-bridge cycle for muscle II. Steady-state behavior. *Biophys. J.* 69: 538–552, 1995.

[87] T. A. J. Duke. Molecular model of muscle contraction. *Proc. Natl. Acad. Sci. USA* 96: 2770–2775, 1999.

[88] A. V. Hill. The heat of shortening and the dynamic constants of muscle. *Proc. R. Soc. Lond.* B 126: 136–195, 1938.

[89] K. A. P. Edman and J. C. Hwang. The force–velocity relationship in vertebrate muscle fibres at varied tonicity of the extracellular medium. *J. Physiol.* 269: 255–272, 1977.

[90] R. Cooke and W. Bialek. Contraction of glycerinated muscle fibers as a function of the MgATP concentration. *Biophys. J.* 28: 241–258, 1979.

[91] R. Cooke and E. Pate. The effect of ADP and phosphate on the contraction of muscle fibers. *Biophys. J.* 48: 789–798, 1985.

[92] Y. Harada, A. Noguchi, A. Kishino, and T. Yanagida. Sliding movement of single actin filaments on one-headed myosin filaments. *Nature* 326: 805–808, 1987.

[93] E. Pate, M. Lin, K. Franks-Skiba, and R. Cooke. Contraction of glycerinated rabbit slow-twitch muscle fibers as a function of MgATP contraction. *Am. J. Physiol.* 262: C1039–C1046, 1992.

[94] B. Katz. The relation between force and speed in muscle contraction. *J. Physiol.* 96: 45–64, 1939.

[95] J. Sleep and H. Glyn. Inhibition of myofibrillar and actomyosin subfragment 1 adenosinetriphosphatase by adenosine 5'-diphosphate and adenyl-5'-yl imidodiphosphate. *Biochemistry* 25: 1149–1154, 1986.

[96] A. F. Huxley and R. M. Simmons. Proposed mechanism of force generation in striated muscle. *Nature* 233: 533–538, 1971.

[97] A. Ishijima, H. Kojima, T. Funatsu, M. Tokunaga, H. Higuchi, H. Tanaka, and T. Yanagida. Simultaneous observation of individual ATPase and mechanical events by a single myosin molecule during interaction with actin. *Cell* 92: 161–171, 1998.

[98] C. Veigel, L. M. Coluccio, J. D. Jontes, J. C. Sparrow, R. A. Milligan, and J. E. Molloy. The motor protein myosin-I produces its working stroke in two steps. *Nature* 398: 530–533, 1999.

[99] K. Kitamura, M. Tokunaga, A. H. Iwane, and T. Yanagida. A single myosin head moves along an actin filament with regular steps of 5.3 nanometres. *Nature* 397: 129–134, 1999.

[100] I. Derényi and T. Vicsek. Unified quantitative theory for the ATP synthase. *Europhys. Lett.* 1999. (submitted).

[101] T. Elston, H. Wang, and G. Oster. Energy transduction in ATP synthase. *Nature* 391: 510–513, 1998.

[102] H. Wang and G. Oster. Energy transduction in the F1 motor of ATP synthase. *Nature* 396: 279–282, 1999.

[103] P. Dimroth, H. Wang, M. Grabe, and G. Oster. Energy transduction in the sodium F-ATPase of propionigenium modestum. *Proc. Natl. Acad. Sci. USA* 96: 4924–4929, 1999.

[104] R. Yasuda, H. Noji, K. Kinosita, and M. Yoshida. F_1-ATPase is a highly efficient molecular motor that rotates with discrete 120° steps. *Cell* 93: 1117–1124, 1998.

[105] I. Derényi, P. Tegzes, and T. Vicsek. Collective transport in locally asymmetric periodic structures. *Chaos* 8: 657–664, 1998.

[106] I. Derényi and T. Vicsek. Cooperative transport of Brownian particles. *Phys. Rev. Lett.* 75: 374–377, 995.

[107] I. Derényi and A. Ajdari. Collective transport of particles in a 'flashing' periodic potential. *Phys. Rev. E* 54: R5–R8, 1996.

[108] F. Jülicher and J. Prost. Cooperative molecular motors. *Phys. Rev. Lett.* 75: 2618–2621, 1995.

[109] Z. Csahók, F. Family, and T. Vicsek. Transport of elastically coupled particles in an asymmetric periodic potential. *Phys. Rev. E* 55: 5179–5183, 1997.

[110] M. R. Evans, D. P. Foster, C. Godrèche, and D. Mukamel. Spontaneous symmetry breaking in a one-dimensional driven diffusive system. *Phys. Rev. Lett.* 74: 208–211, 1995.

[111] F. Jülicher and J. Prost. Spontaneous oscillations of collective molecular motors. *Phys. Rev. Lett.* 78: 4510–4513, 1997.

6 Collective motion

6.1 Flocking: collective motion of self-propelled particles

A. Czirók and T. Vicsek

Introduction

The collective motion of various organisms like the flocking of birds (Fig. 6.1) [1], the swimming of schools of fish [2], and the motion of herds of quadrupeds, migrating bacteria (§4.1), ants [3] or pedestrians (§6.2) is a fascinating phenomenon catching our eyes when we observe our natural environment. In addition to the aesthetic aspects, studies on this collective behaviour can have interesting applications as well, e.g., modelling the motion of a crowd of people can help urban designers. Here we address the question whether there are some global, perhaps universal, features of this type of behaviour when many organisms are involved and parameters like the level of *perturbations* or the mean *distance* between the individuals are changed.

These studies are also motivated by recent developments in areas related to statistical physics. Concepts originating from the physics of phase transitions in equilibrium systems [4,5] such as scale invariance and renormalization have also been shown to be useful in the understanding of various nonequilibrium systems, typical in our natural

Fig. 6.1 Flocking of birds as a typical manifestation of collective motion in biology.

and social environment. Motion and related transport phenomena represent characteristic aspects of many nonequilibrium processes and they are essential features of most living systems.

To understand the collective motion of large groups of organisms the concept of *self-propelled particle* (SPP) models was introduced in [6]. As the motion of flocking organisms is usually controlled by interactions with their neighbours, the SPP models consist of locally interacting particles with an intrinsic driving force and, hence, with a finite steady velocity. Because of their simplicity, such models represent a statistical approach that complements other studies which take into account more details of the actual behaviour [1, 7, 8], but treat only a moderate number of organisms and concentrate less on the large-scale behaviour.

In spite of the analogies with *ferromagnetic* models, the general behaviour of SPP systems can be quite different from those observed in equilibrium magnets. In particular, equilibrium ferromagnets possessing continuous rotational symmetry do not have ordered phase at finite temperatures in two dimensions [9]. However, in 2D SPP models an ordered phase can exist at finite noise levels (temperatures), as was first demonstrated by simulations [6, 10] and explained by a theory of flocking developed by Toner and Tu [11]. Further studies revealed that modelling collective motion leads to interesting specific results in all of the relevant dimensions (from 1 to 3).

6.1.1 Models and simulations

A generic model for two-dimensional SPP systems

The simplest model, introduced in [6], consists of N particles moving in a plane with periodic boundary condition. The particles are characterized by their (off-lattice) location x_i and velocity v_i pointing in direction ϑ_i. The self-propelled nature of the particles is manifested by keeping the magnitude of the velocity fixed to v_0. Particles interact through the following local rule: at each time-step a given particle assumes the *average direction of motion* of the particles in its local neighbourhood $S(i)$ (e.g., in a circle of some given radius centred at the position of the ith particle) with some uncertainty, as described by

$$\vartheta_i(t + \Delta t) = \langle \vartheta(t) \rangle_{S(i)} + \xi, \tag{6.1}$$

where the noise ξ is a random variable with a uniform distribution in the interval $[-\eta/2, \eta/2]$. The locations of the particles are updated as

$$x_i(t + \Delta t) = x_i(t) + v_i(t)\Delta t \tag{6.2}$$

with $|v_i| = v_0 = $ const; see Fig. 6.2. The radius of interaction is the natural unit length in this model to which the system's linear size L and its number density are related to.

The model defined by eqns (6.1) and (6.2) is a transport-related, nonequilibrium analogue of the ferromagnetic models [12]. The analogy is as follows: the Hamiltonian tending to align the spins in the same direction in the case of equilibrium

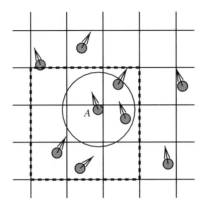

Fig. 6.2 Schematic illustration of the model. Particles move off-lattice on a plane and interact with other particles located in the local surrounding, which can be either a circle or 9 neighbouring cells in an underlying lattice. We plot these interaction areas for particle A with a solid and dashed line, respectively.

ferromagnets is replaced by the rule of aligning the direction of motion of particles, and the amplitude of the random perturbations can be considered proportional to the temperature for $\eta \ll 1$. From a hydrodynamical point of view, in SPP systems the momentum of the particles is *not* conserved. Thus, the flow field emerging in these models can differ considerably from the usual behaviour of fluids.

6.1.2 Scaling properties

The model defined through eqns (6.1) and (6.2) was studied by performing large-scale Monte Carlo simulations in [10]. Due to the simplicity of the model, only two control parameters should be distinguished: the (average) density of particles ϱ and the amplitude of the noise η. Depending on the value of these parameters the model can exhibit various type of behaviour, as Fig. 6.3 demonstrates.

For the statistical characterization of the system a well-suited order parameter is the magnitude of the *average momentum* of the particles:

$$\phi \equiv \frac{1}{N} \left| \sum_j v_j \right|. \tag{6.3}$$

This measure of the net flow is non-zero in the ordered phase, and vanishes (for an infinite system) in the disordered phase. Since the simulations were started from a random, disordered configuration, $\phi(t = 0) \approx 0$. After some relaxation time a steady state emerges, indicated by the convergence of the cumulative average $(1/\tau) \int_0^\tau \phi(t) \, dt$. The stationary values of ϕ versus η are plotted in Fig. 6.4(a) for $\varrho = 2$ and various system sizes L. For weak noise the model displays an ordered motion, i.e. $\phi \approx 1$, which disappears in a continuous manner by increasing η.

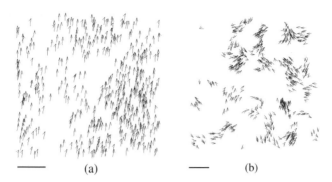

Fig. 6.3 Typical configurations of SPPs displayed for various values of density and noise. The actual velocity of a particle is indicated by a small arrow, while its trajectory for the last 20 time-steps is shown by a short continuous curve. For comparison, the radius of the interaction is displayed as a bar. (a) At high densities and small noise ($N = 300$, $L = 5$ and $\eta = 0.1$) the motion becomes ordered. (b) For small densities and noise ($N = 300$, $L = 25$ and $\eta = 0.1$) the particles tend to form groups moving coherently in random directions (after [6]). Copyright 1995 by the American Physical Society.

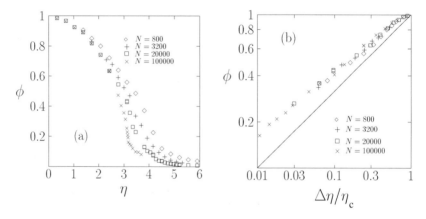

Fig. 6.4 (a) The average momentum of the 2D SPP model in the steady state versus the noise amplitude η for $\varrho = 2$ and four different system sizes [($\Diamond$) $N = 800$, $L = 20$; ($+$) $N = 3200$, $L = 40$; ($\Box$) $N = 20000$, $L = 100$ and ($\times$) $N = 10^5$, $L = 223$]. (b) The order present at small η disappears in a continuous manner reminiscent of second-order phase transitions: $\phi \sim [(\eta_c(L) - \eta)/\eta_c(L)]^\beta \equiv (\Delta\eta/\eta_c)^\beta$, with $\beta = 0.42$, different from the mean-field value $1/2$ (solid line) (after [10]).

As $L \rightarrow \infty$, the numerical results show the presence of a phase transition described by

$$\phi(\eta) \sim \begin{cases} ((\eta_c(\varrho) - \eta)/(\eta_c(\varrho)))^\beta & \text{for } \eta < \eta_c(\varrho) \\ 0 & \text{for } \eta > \eta_c(\varrho), \end{cases} \quad (6.4)$$

where $\eta_c(\varrho)$ is the critical noise amplitude that separates the ordered and disordered phases and

$$\beta_{2d} = 0.42 \pm 0.03, \quad (6.5)$$

which is *different* from the the mean-field value $\frac{1}{2}$ (Fig 6.4(b)) and *insensitive* to the specific choice of the interaction range $S(i)$.

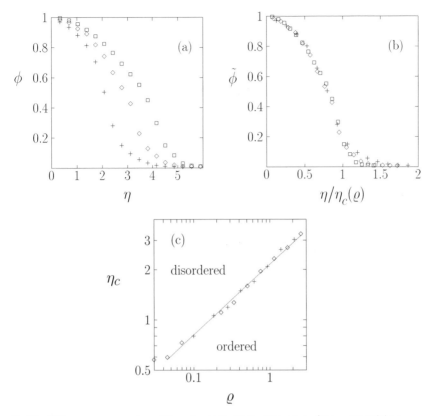

Fig. 6.5 (a) The average momentum of the system in the steady state versus η for $L = 100$ and three different densities [($\square$) $\varrho = 4$, ($\lozenge$) $\varrho = 2$ and (+) $\varrho = 0.5$]. (b) The $\phi(\eta)$ functions parametrized by various ϱ can be collapsed onto a single curve $\tilde{\phi}[\eta/\eta_c(\varrho)] \equiv \phi(\eta, \varrho)$. (c) The critical line in the $\eta - \varrho$ phase space is a power law in the examined regime: $\eta_c(\varrho) \sim \varrho^\kappa$ with $\kappa \approx 0.45$ (solid line) for a system of size $L = 100$ (after [10]).

Next we discuss the role of density. In Fig. 6.5(a), $\phi(\eta)$ is plotted for $L = 100$ and various values of ϱ. One can observe that the long-range ordered phase is present for any ϱ, but for a fixed value of η, ϕ vanishes with decreasing ϱ. All of the $\phi(\eta, \varrho)$ functions can be collapsed to the same function $\tilde{\phi}(x)$ by rescaling η with $\eta_c(\varrho)$:

$$\phi(\eta, \varrho) = \tilde{\phi}(\eta/\eta_c(\varrho)), \tag{6.6}$$

where $\tilde{\phi}(x) \sim (1 - x)^\beta$ for $x < 1$, and $\tilde{\phi}(x) \approx 0$ for $x > 1$ (see Fig. 6.5(b)). The critical line $\eta_c(\varrho)$ in the $\eta - \varrho$ parameter space was found to follow

$$\eta_c(\varrho) \sim \varrho^\kappa, \tag{6.7}$$

with $\kappa = 0.45 \pm 0.05$ (see Fig. 6.5(c)). This critical line is qualitatively different from that of the diluted ferromagnets, since here the critical density as $\eta \to 0$ (corresponding to the percolation threshold for diluted ferromagnets, e.g., see [12]) vanishes to zero.

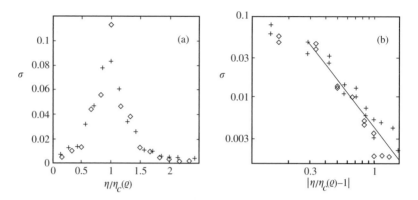

Fig. 6.6 (a) The rms deviation of the order parameter σ in the steady state, for $L = 100$, and two independent data sets ($\Diamond$, $+$), each averaged over runs with $\varrho = 0.3, 0.6, 0.9, 1.2, 1.5$ and 2. (b) The divergence is symmetric and its tail decays as $\sigma(x) \sim |1 - x|^{-2}$ (solid line), where x denotes the rescaled noise amplitude $\eta/\eta_c(\varrho)$ (after [10]).

Note, that eqn (6.6) also implies that the exponent β', defined as $\phi \sim (\varrho - \varrho_c)^{\beta'}$ for $\varrho > \varrho_c$ [6], must be equal to β, since

$$\phi(\eta, \varrho_c(\eta) + \epsilon) = \tilde{\phi}\left(\frac{\eta}{\eta_c[\varrho_c(\eta) + \epsilon]}\right) \approx \tilde{\phi}\left(1 - \frac{1}{\eta}\frac{\partial \eta_c}{\partial \varrho}\bigg|_{\varrho_c}\epsilon\right) \sim \epsilon^{\beta}, \qquad (6.8)$$

where $\eta_c(\varrho)$ denotes the inverse function of $\varrho_c(\eta)$ as $\eta \equiv \eta_c[\varrho_c(\eta)]$.

As a further analogy with equilibrium phase transitions, the fluctuations of the order parameter also increase on approaching the critical line. Fig. 6.6(a) shows the standard deviation σ of the total momentum versus the rescaled noise amplitude $x \equiv \eta/\eta_c(\varrho)$ for various densities and $L = 100$. The tails of the curves are symmetric, and decay as power laws (see Fig. 6.6(b)),

$$\sigma(x) \sim |1 - x|^{-\gamma}, \qquad (6.9)$$

with an exponent γ close to 2, whose value is, again, different from the mean-field result.

These findings indicate that SPP systems can be quite well characterized using the framework of classical critical phenomena, but also show surprising features when compared to the analogous equilibrium systems. The velocity v_0 provides a control parameter which switches between the SPP behaviour ($v_0 > 0$) and an XY ferromagnet ($v_0 = 0$). Indeed, for $v_0 = 0$ Kosterlitz–Thouless vortices [13] can be observed in the system, which are *unstable* (Fig. 6.7) for any nonzero v_0 investigated in [10].

3D SPP system

In two dimensions an effective long-range interaction can build up because the migrating particles have a considerably higher chance to get close to each other and interact

Fig. 6.7 Snapshots of the time development of a system with $N = 4000$, $L = 40$ and $v_0 = 0.01$ at (a) 50, (b) 100, (c) 400 and (d) 3000 Monte Carlo steps. At first, the behaviour is reminiscent of the equilibrium XY model, where the long-range order is missing since vortices are present in the system. Here the vortices are unstable, and finally a self-organized long-range order develops (after [10]).

than in three dimensions (where, as is well known, random trajectories do not overlap). The less interaction acts against ordering. On the other hand, in three dimensions even regular ferromagnets order. Thus, it is interesting to see how these two competing features change the behaviour of 3D SPP systems. The convenient generalization of eqn (6.1) for the 3D case can be the following [14]:

$$\mathbf{v}_i(t + \Delta t) = v_0 \mathbf{N}(\mathbf{N}(\langle \mathbf{v}(t) \rangle_{S(i)}) + \boldsymbol{\xi}), \tag{6.10}$$

where $\mathbf{N}(\mathbf{u}) = \mathbf{u}/|\mathbf{u}|$ and the noise $\boldsymbol{\xi}$ is uniformly distributed in a sphere of radius η.

Generally, the behaviour of the system was found [14] to be similar to that described in the previous section. The long-range ordered phase was present for any ϱ, but for a fixed value of η, ϕ vanished with decreasing ϱ. To compare this behaviour to the corresponding diluted ferromagnet, $\phi(\eta, \varrho)$ was determined for $v_0 = 0$, when the model reduces to an equilibrium system of randomly distributed 'spins' with a ferromagnetic-like interaction. Again, a major difference was found between the SPP and the equilibrium models (Figs. 6.8 and 6.9): in the static case the system *does not order* for densities below a critical value close to 1 which corresponds to the percolation threshold of randomly distributed spheres in 3D.

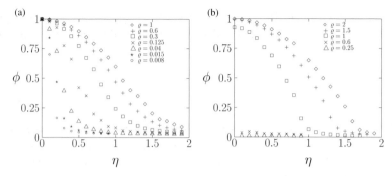

Fig. 6.8 (a) The order parameter ϕ versus the noise amplitude η ($N = 1000$) for the 3D SPP system. (b) As a comparison, when $v_0 = 0$ the behaviour of the model is similar to diluted ferromagnets: ϕ vanishes below the percolation threshold ($\rho^* \simeq 1$) (after [14]). Copyright 1999 reprinted with permission from Elsevier Science.

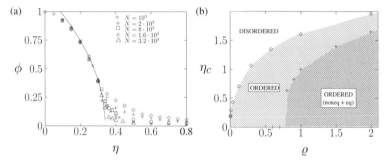

Fig. 6.9 (a) The order parameter ϕ versus the noise amplitude η for the 3D SPP model for various system sizes. In these simulations the density was fixed and the system size (number of particles N) was increased to demonstrate that for any system size the ordered phase disappears in a continuous manner beyond a size-dependent critical noise. (b) Phase diagram of the 3D SPP and the corresponding ferromagnetic system. The diamonds show our estimates for the critical noise for a given density for the SPP model and the crosses show the same for the static case. The SPP system becomes ordered in the whole region below the curved line connecting the diamonds, while in the static case the ordered region extends only down to the percolation threshold $\rho^* \simeq 1$ (after [14]). Copyright 1999 reprinted with permission from Elsevier Science.

1D SPP system

Since in 1D the particles cannot get around each other, some of the important features of the dynamics present in higher dimensions are lost. On the other hand, motion in 1D implies new interesting aspects (groups of the particles have to be able to change their direction to the opposite in an organized manner) and the algorithms used for higher dimensions should be modified to take into account the specific crowding effects typical for 1D (the particles can slow down before changing direction and dense regions may be built up of momentarily oppositely moving particles).

In a way the system studied below can be considered as a model of people moving in a narrow channel. Imagine that a fire alarm goes on, the tunnel is dark and smoky, and everyone is extremely excited. People are both trying to follow the others (to escape together) and behaving in an erratic manner (due to smoke and excitement).

Thus, in [15] N off-lattice particles along a line of length L have been considered. The particles are characterized by their coordinate x_i and dimensionless velocity u_i updated as

$$x_i(t + \Delta t) = x_i(t) + v_0 u_i(t) \Delta t, \qquad (6.11)$$

$$u_i(t + \Delta t) = G\big(\langle u(t) \rangle_{S(i)}\big) + \xi_i. \qquad (6.12)$$

The local average velocity $\langle u \rangle_{S(i)}$ for the ith particle is calculated over the particles located in the interval $[x_i - \Delta, x_i + \Delta]$, where $\Delta = 1/2$. The function G incorporates both the 'propulsion' and 'friction' forces which set the velocity to a prescribed value v_0 on the average: $G(u) > u$ for $u < 1$ and $G(u) < u$ for $u > 1$. In the numerical simulations [15] one of the simplest choices for G was implemented as

$$G(u) = \begin{cases} (u+1)/2 & \text{for } u > 0 \\ (u-1)/2 & \text{for } u < 0, \end{cases} \qquad (6.13)$$

and random initial and periodic boundary conditions were applied.

In Fig. 6.10 we show the time evolution of the model for $\eta = 2.0$. In a short time the system reaches an ordered state, characterized by a spontaneous broken symmetry and clustering of the particles. In contrast, for $\eta = 4.0$ the system remains in a disordered state.

The order parameter for various values of ϱ is shown in Fig. 6.11. As in the two- and three-dimensional cases, the emergence of the ordered phase was observed through a second-order phase transition (Fig. 6.11(b)) with

$$\beta_{1d} = 0.60 \pm 0.05, \qquad (6.14)$$

which is different from both the the mean-field value $\frac{1}{2}$ and $\beta_{2d} \approx 0.4$ found in 2D. Figure 6.11(c) shows that the various $\phi(\eta, \varrho)$ curves can be collapsed onto a single

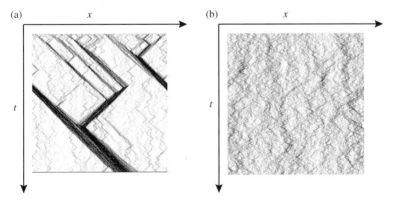

(a) x (b) x

t

Fig. 6.10 The first 3000 time-steps of the 1D SPP model ($L = 300$, $N = 600$, $\eta = 2.0$ (a) and $\eta = 4.0$ (b)). The darker grey scale represents higher particle density (after [15]). Copyright 1999 by the American Physical Society.

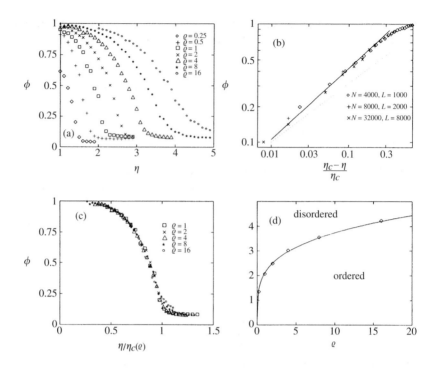

Fig. 6.11 (a) The order parameter ϕ versus η for the 1D SPP model ($L = 1000$). (b) ϕ vanishes as a power-law in the vicinity of $\eta_c(\rho)$. Note the increasing scaling regime with increasing L. The solid line is a power-law fit with an exponent $\beta = 0.6$, while the dotted line shows the mean-field slope $\beta = 1/2$ as a comparison. (c) As in the 2D case, the $\phi(\eta, \varrho)$ functions can be collapsed onto a single curve. (d) Phase diagram in the $\rho - \eta$ plane; the critical line follows $\eta_c(\rho) \sim \rho^{\kappa}$. The solid curve represents a fit with $\kappa = 1/4$ (after [15]). Copyright 1999 by the American Physical Society.

function $\phi_0(x)$, where $x = \eta/\eta_c(\varrho)$, just as in 2D. The $\rho - \eta$ phase diagram is shown in Fig. 6.11(d). The critical line, $\eta_c(\varrho)$, follows eqn (6.7) with

$$\kappa_{1d} \approx \tfrac{1}{4}. \tag{6.15}$$

6.1.3 Further variants of SPP models

Effect of boundary conditions.

In various simulations [16–18] the boundary conditions were changed to *reflective circular walls* to gain insight into the importance of the boundary conditions in the SPP models. To avoid singular behaviour at the boundaries, i.e., the aggregation of particles in a narrow zone, a *short-range 'hardcore' repulsion* was also incorporated into the models: E.g., in the simulations described in [16], if the distance between the particles is smaller than a certain ϵ^* value they repel each other and the direction of their motion will be given by the following expression instead of eqn (6.1):

$$\vartheta_i(t + \Delta t) = \Phi\left(-\sum_{\substack{j \neq i \\ |x_j - x_i| < \epsilon^*}} N(x_j(t) - x_i(t))\right),$$ (6.16)

as demonstrated in Fig. 6.12.

In such simulations rotation of the particles develop (see Fig. 6.13) in the high-density, low-noise regime. The direction of rotation is selected by spontaneous symmetry breaking; thus both clockwise and anti-clockwise spinning 'vortices' can emerge. This rotating state should be distinguished from the Kosterlitz–Thouless vortices [13], since in our case a *single* vortex develops irrespective of the system size. In §4.1 we give examples of such vortices developing in nature.

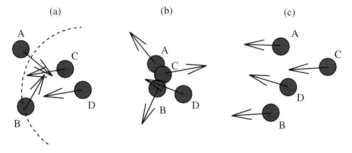

Fig. 6.12 Schematic illustration of the model with both hardcore interaction and ferromagnetic-like coupling of the velocities. The figure shows the position and velocity (arrows) of the particles at time (a) t, (b) $t + \Delta t$ and (c) $t + 2\Delta t$. The range of the hardcore repulsion (ϵ^*) is represented as a filled circle. The dashed circle in (a) indicates the range of velocity interaction for particle D. In the collision at time $t + \Delta t$ particles A, B and C repel each other, while the new velocity of particle D was determined by the average direction of motion of B, C and D. At time $t + 2\Delta t$ no collision occurs, and each particle moves approximately with the local average velocity (after [16]). Copyright 1996 by the American Physical Society.

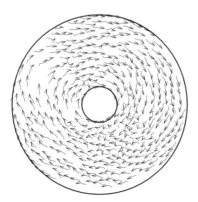

Fig. 6.13 A possible stationary state of the model represented in Fig. 6.12 with reflective boundary conditions (after [16]). Copyright 1996 by the American Physical Society.

Lattice gas approach

As discrete cellular automata models often have the advantages that they allow both an approximate analytic and large-scale numerical investigations, in [19] the following discrete model for SPP systems was introduced and studied: each site of a triangular lattice can be either empty or occupied by one or more particles. If more than one particle is present at a site then only one of them, called the 'lowest', will interact with other particles. The particles are characterized by their position r_i and velocity v_i ($i = 1, \ldots, N$) which is of unit length ($|v_i| = 1$) and can point in any of the six lattice directions (u_α). At one time-step the positions and velocities of all particles are updated simultaneously according to the following rules:

1. (a) For noninteracting particles a random direction is assigned.

   (b) Otherwise, a new velocity u_α is chosen with a Boltzmann distribution as

$$P(u_\alpha) = \frac{1}{\mathcal{Z}} \exp\left(-\beta u_\alpha \sum_{j \in \mathrm{lnn}} v_j\right), \tag{6.17}$$

where $\mathcal{Z}$ is a normalizing factor such that $\sum_{\alpha=1}^{6} P(u_\alpha) = 1$, and β is $1/T$ ($k_B = 1$). Thus, in this case there is a direct correspondence between the noise strength of the former models and some temperature T. The summation goes over the nearest neighbours which are in lowest position (lnn). This rule is equivalent to having an equilibrium spin system with the Hamiltonian

$$H = -\frac{1}{2} \sum_{i,j \in \mathrm{lnn}} v_i v_j. \tag{6.18}$$

2. Every particle is moved one lattice unit in direction of its velocity:

$$r_i \leftarrow r_i + v_i.$$

This system is closely related to the 6-state Potts model [20], since we have $q = 6$ possible velocity states for a particle, but unlike in the Potts model these states are not orthogonal and the system is not in equilibrium.

As the initial configuration, a random distribution for the position and velocity of the particles was chosen in [19]. Typical configurations of the system for various temperatures T and particle densities ϱ are shown in Fig. 6.14. In a similar fashion to the behaviour of the corresponding models with rotational symmetry, it can be seen that at low temperatures the particles tend to form ordered clusters.

Figure 6.15(a) demonstrates the order parameter as a function of the temperature for $\varrho = 0.9$. Note the discontinuity that is characteristic of first-order transitions at T_c. Strong evidence supporting the presence of a first-order transition is shown in Fig. 6.15(b), where the distribution $P(\varepsilon)$ of the energy of the particles is plotted for a number of different temperatures below and above T_c. One can clearly see a *gap* in the distributions at intermediate energies, which is a unique feature of first-order

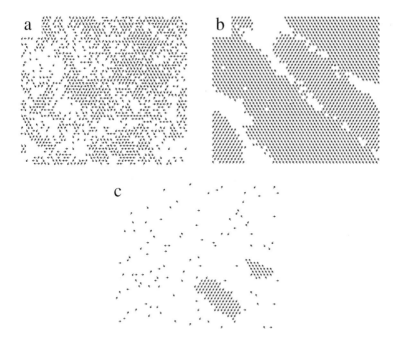

Fig. 6.14 Typical snapshots of the system at (a) high temperature, (b) low temperature at $\varrho = 0.9$ and (c) intermediate temperature at a lower density. Note the appearance of ordered clusters. (Only the particles in the lowest position are drawn) (after [19]). Copyright 1995 by the American Physical Society.

phase transitions [21]. The inset in the figure shows the distribution for Ising type interaction of nonmoving particles in the same system where the transition is known to be second order. This transition was found to be present even for very small particle densities although the position of the critical temperature decreases with the density just as in the case of the off-lattice models.

SPP model in SOC state

In [22] another variant of the 2D model was investigated where the noise amplitude was dependent on the local particle density ρ_ℓ as

$$\eta = \pi \rho_\ell^{-\alpha}, \tag{6.19}$$

instead of being an external control parameter. In addition to the eqns (6.1) and (6.2) the number of particles were determined by local population dynamics rules.

The angular correlation plot is shown in Fig. 6.16: for $\alpha \to 0$ there are many flocks in the system moving independently, while for large α values all the particles are organized into a single flock. The transition was found to be continuous between the two regimes.

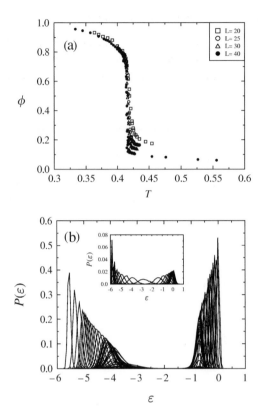

Fig. 6.15 Statistical characterization of the transition between ordered and disordered systems. (a) the order parameter (ϕ) as a function of the temperature (T) for density $\varrho = 0.9$. (b) the energy distribution ($P(\varepsilon)$) for various temperatures below and above the transition ($\varrho = 0.9$). Note the energy gap between $\varepsilon \approx -1$ and $\varepsilon \approx -3.5$. Inset shows the distribution for Ising spins instead of mobile particles where the transition is continuous (after [19]). Copyright 1995 by the American Physical Society.

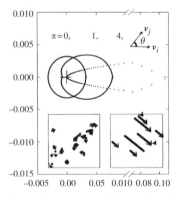

Fig. 6.16 Polar plot of the angular correlation function of the velocities of the migrating individuals for three values of α. The inserts below show configurations with $\alpha = 0$ (left) with random migration and $\alpha = 4$ (right) where ordered motion develops (after [22]). Copyright 1996 by the American Physical Society.

Deterministic model for the motion of organisms

Many attempts have been made to describe the actual behaviour of various organisms using effective interaction potentials or forces. In [7] a deterministic model was introduced describing the dynamics of the individuals in terms of their velocity, position, and a *heading unit vector* $\boldsymbol{n}$ as

$$m\dot{\boldsymbol{v}}_i = -\gamma \boldsymbol{v}_i + a\boldsymbol{n}_i + \sum_{i \neq j} \alpha_{ij} \boldsymbol{f}_{ij} + \boldsymbol{g}_i \qquad (6.20)$$

$$\tau \dot{\boldsymbol{n}}_i = \boldsymbol{n}_i \times \mathrm{N}(\boldsymbol{v}_i) \times \boldsymbol{n}_i \qquad (6.21)$$

where eqn (6.20) is Newton's law of motion for particles of mass m; γ is the damping coefficient based on Stokes' law. The locomotive force of magnitude a acts in the heading direction $\boldsymbol{n}$. The $\boldsymbol{f}_{ij}$ force between particles j and i incorporates effects providing a mechanical equilibrium at a given distance r_0, and in general, $\boldsymbol{f}_{ij} \neq \boldsymbol{f}_{ji}$. The term $\boldsymbol{g}_i$ represent a force towards the centre of the group. $\boldsymbol{v}$ and $\boldsymbol{n}$ are assumed to relax together with relaxation time τ according to eqn (6.21).

The model is fairly complex and exhibits several distinct types of collective behaviour (Fig. 6.17). For example, a continuous transition was found as a function of the corresponding dimensionless parameters between a disordered (chaotic) and an ordered (nonchaotic) phase, resembling the transition found in the stochastic 2D SPP model.

Theoretical results

To understand the properties of the ordered phase or the phase transitions observed in the models described in the previous section, efforts have been made to set up a consistent continuum theory in terms of $\mathbf{v}$ and ρ, representing the coarse-grained velocity and density fields, respectively. The first such approach was made by J. Toner

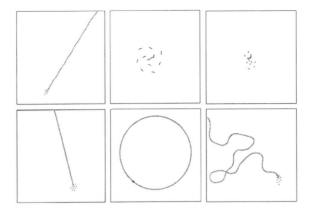

Fig. 6.17 Patterns of clusters and trajectories of their centres for various values of the control parameters. Solid bars represent the motive elements, and the grey line is the trajectory of the centre of the cluster (after [7]). Copyright 1996 by the American Physical Society.

and Y. Tu [11]. Here, however, we overview the efforts not in chronological order, but in a progressive fashion starting from the most simple mean-field lattice theory through the one-dimensional system to the most difficult higher dimensional cases.

6.1.4 Mean-field theory for lattice models

Equilibrium mean-field theory

As we discussed in the previous section, in [19] a cellular automata SPP model was introduced and studied. The major advantage of such cellular automata models is the possibility of an approximate analytic investigation which, as we demonstrate here following [19], often gives surprisingly good numerical results and provides a general understanding of the system.

As the first step of building a mean-field theory, a *mean-field Hamiltonian* can be introduced [23] instead of eqn (6.18) as

$$H_{MF} = -\frac{1}{2} \sum_{i,j\in l} \boldsymbol{v}_i \boldsymbol{v}_j,$$ (6.22)

where the summation goes over *all lowest* particles, not only the nearest neighbours. The mean-field energy function can be written as

$$E_{MF} = -\frac{1}{2} N \varrho_{eff} \sum_{\alpha,\gamma=1}^{6} x_\alpha U_{\alpha\gamma} x_\gamma,$$ (6.23)

where $\varrho_{eff} = 1 - \exp(-\varrho)$ is the effective density, i.e., a site has on average $6\varrho_{eff}$ occupied neighbouring sites, x_α is the fraction of particles travelling in the lattice direction α ($\sum_{\alpha=1}^{6} x_\alpha = 1$) and $U_{\alpha\gamma} = \boldsymbol{u}_\alpha \boldsymbol{u}_\gamma$, which for the case of the Potts model would be simply $U_{\alpha\gamma} = \delta_{\alpha\gamma}$. Then, the average energy per particle is

$$\varepsilon_{MF} = \frac{E_{MF}}{N} = -\frac{1}{2} \varrho_{eff} \sum_{\alpha,\gamma=1}^{6} x_\alpha U_{\alpha\gamma} x_\gamma,$$ (6.24)

while the entropy per particle is

$$s_{MF} = -\sum_{\alpha=1}^{6} x_\alpha \ln x_\alpha.$$ (6.25)

So for the free energy per particle we obtain

$$\beta f_{MF} = \beta \frac{F_{MF}}{N} = \beta \varepsilon_{MF} - s_{MF} = \sum_{\alpha=1}^{6} \left(x_\alpha \ln x_\alpha - \frac{1}{2} \varrho_{eff} \beta x_\alpha \sum_{\gamma=1}^{6} U_{\alpha\gamma} x_\gamma \right).$$ (6.26)

We intend to find the configuration x_α which minimizes the free energy f_{MF}. Since all the lattice directions are equivalent, we can look for a solution in the form of

$$x_1 = \tfrac{1}{6} + \tfrac{5}{6}\phi_{MF}$$ (6.27)

and

$$X_{\alpha>1} = \frac{1}{6} - \frac{\phi_{MF}}{6},\tag{6.28}$$

where ϕ_{MF} is the mean-field order parameter which satisfies the relation

$$\phi_{MF} = \left| \sum_{\alpha=1}^{6} x_\alpha \mathbf{u}_\alpha \right|.$$

Substituting eqns (6.27) and (6.28) into eqn (6.26), after a bit of algebra one gets for the mean-field free energy

$$\beta f_{MF} = -\frac{1}{2}\varrho_{eff}\beta \, \phi_{MF}^2 - \log\frac{1}{6} + \frac{5(1-\phi_{MF})}{6}\log\frac{1-\phi_{MF}}{6}$$
$$+ \frac{1+5\phi_{MF}}{6}\log\frac{1+5\phi_{MF}}{6}.\tag{6.29}$$

For high temperatures $(T > T_c)$ this function has its minimum at $\phi_{MF} = 0$, which corresponds to the disordered state of the system. At the critical temperature, which can be derived from f_{MF} and in our case is

$$T_c \approx \frac{\varrho_{eff}}{3.353},\tag{6.30}$$

a nontrivial minimum appears as demonstrated in Fig. 6.18. The phase transition, as in the 6-state Potts model [24], is first order. The jump in the order parameter in this approximation is given exactly by

$$\Delta\phi_{MF} = 0.8,$$

which is in reasonable agreement with the numerical findings presented in Fig. 6.15(a).

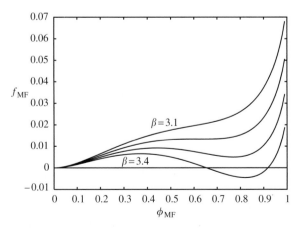

Fig. 6.18 The mean-field free energy f_{MF} as given by eqn (6.29) versus the order parameter ϕ_{MF} for $\varrho_{eff} = 1$ and for various β values in the vicinity of the phase transition.

Nonequilibrium mean-field theory

It is also possible to carry out a mean-field analysis of the problem without assuming thermal equilibrium, as was first described in [25]. The specific model we consider here is slightly different from the one studied in the previous section: each node r can contain up to four particles in different 'velocity channels' corresponding to nearest neighbour sites. The state of the entire lattice at time t is specified by the occupation numbers $s_i(r, t) = 0, 1$ denoting the absence or presence of a particle in the channel c_i at node r.

Similarly to the previous lattice model, the evolution from time t to $t+1$ proceeds in two stages: first an interaction step is performed during which the preinteraction states $\{s_i(r, t)\}$ are replaced by a postinteraction state $\{\sigma_i(r, t)\}$ according to stochastic rules that are applied to each node r independently. The interaction step is then followed by a propagation step during which particles move to nearest neighbour sites in the direction of their velocity as

$$s_i(r + c_i, t + 1) = \sigma_i(r, t). \tag{6.31}$$

The local alignment interaction is implemented by calculating the average flux of particles at the nearest neighbours

$$D(r, t) = \sum_{p=1}^{4} \sum_{i=1}^{4} c_i s_i(r + c_p, t). \tag{6.32}$$

The number of particles at each node, $\rho(r, t) = \rho[s(r, t)] \equiv \sum_{i=1}^{4} s_i(r, t)$, is conserved during the interaction. Thus, if $J(\sigma) = \sum_{i=1}^{4} c_i \sigma_i$ denotes the particle flux immediately after the interaction, then the transition probability from $s(r, t)$ to $\sigma(r, t)$ in the presence of $D(r, t)$ is given by

$$A[s \to \sigma | D] = \frac{1}{\mathcal{Z}} \delta[\rho(\sigma), \rho(s)] \exp[\beta D \cdot J(\sigma)], \tag{6.33}$$

where the normalization factor $\mathcal{Z}$ is chosen such that $\sum_{\sigma} A[s \to \sigma | D] = 1$ for all s. Note that this interaction rule minimizes the angle between the director field D and the postinteraction flux $J(\sigma)$ in a similar manner to eqn (6.17). Figure 6.19 shows the time evolution of an initially random distribution for $\beta > \beta_c$. The formation of locally aligned patches can clearly be observed.

The behaviour of such models can be analysed by considering the time evolution of a statistical ensemble of systems as discussed in [26]. For the mean-field description the ensemble-averaged occupation numbers $f_i(r, t) \equiv \overline{s_i(r, t)}$ are introduced. It is assumed that at each time-step just before interaction the probability distribution is completely factorized, i.e., the probability $\mathcal{P}(\{s_i(r)\})$ of finding a microstate $\{s_i(r)\}$ at time t is given by

$$\mathcal{P}(\{s_i(r)\}) = \prod_{r} \prod_{i=1}^{4} [f_i(r, t)]^{s_i(r)} [1 - f_i(r, t)]^{1-s_i(r)}. \tag{6.34}$$

$t = 1000$

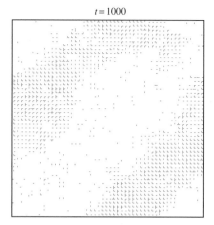

Fig. 6.19 Swarming behaviour in a cellular automaton model. Shown are the occupied velocity channels of the cellular automaton organized into a symmetry-broken state after 1000 time-steps (after [25]). Copyright 1997 by the American Physical Society.

We denote averaging using the above, factorized statistical weights by $\langle \cdots \rangle_{\text{MF}}$. Averaging eqn (6.31) and replacing the real ensemble average by $\langle \cdots \rangle_{\text{MF}}$, i.e., neglecting all correlations between occupation numbers, we obtain a closed evolution equation for $f_i(\boldsymbol{r}, t)$:

$$f_i(\boldsymbol{r} + \boldsymbol{c}_i, t + 1) = f_i(\boldsymbol{r}, t) + I_i(\boldsymbol{r}, t). \tag{6.35}$$

Here the term $I_i(\boldsymbol{r}, t) \equiv \langle \sigma_i(\boldsymbol{r}, t) - s_i(\boldsymbol{r}, t) \rangle_{\text{MF}}$, taking values between -1 and 1, equals the average change in the occupation number of channel $(\boldsymbol{r}, \boldsymbol{c}_i)$ during interaction.

As a consequence of the conservation of particle number, $\sum_i I_i = 0$, combined with the invariance of the interaction rules under discrete rotations and translations, a possible solution to eqn (6.35) is

$$f_i(\boldsymbol{r}, t) = f^* = \rho^*/4. \tag{6.36}$$

To assess the stability of this spatially homogeneous and stationary solution with respect to fluctuations

$$\delta f_i(\boldsymbol{r}, t) = f_i(\boldsymbol{r}, t) - f^*, \tag{6.37}$$

Equation (6.35) can be linearized, resulting in

$$\delta f_i(\boldsymbol{r} + \boldsymbol{c}_i, t + 1) = \delta f_i(\boldsymbol{r}, t) + \sum_{j, \boldsymbol{r}'} \left. \frac{\partial I_i}{\partial f_j(\boldsymbol{r}')} \right|^* \delta f_j(\boldsymbol{r}', t). \tag{6.38}$$

Since the interaction is local, eqn (6.38) can be simplified to

$$\delta f_i(\boldsymbol{r} + \boldsymbol{c}_i, t + 1) = \delta f_i(\boldsymbol{r}, t) + \sum_{j, p} \Omega_{ij}^p \delta f_j(\boldsymbol{r} + \boldsymbol{c}_p, t), \tag{6.39}$$

with $p \in \{0, \ldots, 4\}$, $\boldsymbol{c}_0 = 0$ and

$$\Omega_{ij}^p = \left. \frac{\partial I_i}{\partial f_j(\boldsymbol{r} + \boldsymbol{c}_p)} \right|^*. \tag{6.40}$$

Fourier transforming eqn (6.39) as $\delta \hat{f}_i(\boldsymbol{k}, t) = \sum_r e^{-i\boldsymbol{k}\cdot\boldsymbol{r}} \delta f_i(\boldsymbol{r}, t)$ we obtain

$$\delta \hat{f}_i(\boldsymbol{k}, t+1) \simeq \sum_{j=1}^4 \Gamma_{ij}(\boldsymbol{k}) \delta \hat{f}_j(\boldsymbol{k}, t), \tag{6.41}$$

with

$$\Gamma_{ij}(\boldsymbol{k}) = e^{-i\boldsymbol{k}\cdot\boldsymbol{c}_i} \left[\delta_{ij} + \sum_{p=0}^4 e^{i\boldsymbol{k}\cdot\boldsymbol{c}_p} \, \Omega_{ij}^p \right]. \tag{6.42}$$

It can be shown [25, 26] that $\delta_{ij} + \Omega_{ij}^0 = \frac{1}{4}$ for all i, j; this is a consequence of the fact that the outcome $\sigma(\boldsymbol{r})$ of an interaction step only depends on $s(\boldsymbol{r})$ through $\rho(\boldsymbol{r})$. For $1 \le p \le 4$ the elements $\Omega_{ij}^p \equiv \omega_{ij}$ do not depend on p, and for symmetry reasons $(\omega)_{ij}$ is a cyclic matrix whose first row has the structure $(a+b, -b, -a+b, -b)$. To determine $a(\beta, \rho^*)$ and $b(\beta, \rho^*)$ for given values of the sensitivity β and the average density ρ^*, the expression

$$\omega_{ij} = \sum_{\{s(\boldsymbol{r}+\boldsymbol{c}_p)\}} \sum_{\sigma(\boldsymbol{r})} (\sigma_i(\boldsymbol{r}) - s_i(\boldsymbol{r})) \frac{s_j(\boldsymbol{r}+\boldsymbol{c}_1) - f^*}{f^*(1-f^*)}$$

$$\times \, A[s \to \sigma | \boldsymbol{D}(\{s(\boldsymbol{r}+\boldsymbol{c}_p)\})] \prod_{p'=0}^4 F(s(\boldsymbol{r}+\boldsymbol{c}_{p'})), \tag{6.43}$$

can be evaluated (this is done numerically because of the highly nonlinear dependence on f_i and $\beta \boldsymbol{D}$, combined with the large number of terms) where $F^*(s) = \prod_{i=1}^4 f^{*s_i}(1 - f^*)^{1-s_i}$ is the factorized distribution of the homogeneous equilibrium system.

To investigate the stability of the spatially uniform state, i.e. $\boldsymbol{k} = 0$, note that $\Gamma_{ij}(\boldsymbol{k} = 0)$ has an eigenvalue $\lambda_1 = 1$ with corresponding eigenvector $e_1 = (1, 1, 1, 1)$, reflecting the fact that the total density is conserved. Furthermore there is a twofold-degenerate eigenvalue $\lambda_{x,y} = 8a$ with an eigenspace spanned by $e_x = (1, 0, -1, 0)$ and $e_y = (0, 1, 0, -1)$, corresponding to the x- and y-components of the total particle flux. The remaining eigenvector $e_{x^2-y^2} = (1, -1, 1, -1)$ has eigenvalue $\lambda_{x^2-y^2} = 16b$, corresponding to the difference between the number of horizontally and vertically moving particles. Numerically, b is found [25] to be about two orders of magnitude smaller than a, so that the onset of instability of the homogeneous state is determined by the condition $\lambda_{x,y} = 1$. The location of the critical line in the $(\beta, \bar{\rho})$ parameter plane is shown in Fig. 6.20, which was obtained by numerically solving the equation $a(\beta, \bar{\rho}) = 1/8$. The resulting phase diagram is in qualitative agreement for $\rho^* \sim 1$ with the one shown in Fig. 6.5(c).

6.1.5 Continuum equations for the 1D system

Now let us focus on the continuum approaches. As the simplest case, let us first investigate, following [15], the 1D SPP system described in the previous section. We denote by $n(u, x, t) du \, dx$ the number of particles moving with a velocity in the range

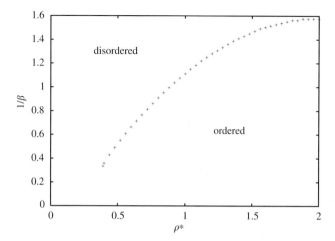

Fig. 6.20 Phase diagram for swarming model. Shown are the regions where the disordered state is stable and unstable against small fluctuations, as a function of 'noise' $1/\beta$ and average density ρ^*. (After [25].)

of $[v_0 u, v_0(u + du)]$ at time t in the $[x, x + dx]$ interval. The particle density $\rho(x, t)$ is then given as

$$\rho = \int n \, du, \tag{6.44}$$

while the local dimensionless average velocity $U(x, t)$ can be calculated as

$$\rho U = \int nu \, du. \tag{6.45}$$

According to the microscopic rule of the dynamics eqn (6.12), in a given time interval $[t, t+\tau]$ all particles choose a certain velocity $v = v_0[G(\langle u \rangle) + \xi]$ and travel a distance $v\tau$. Thus, the time development of the ensemble average (denoted by overline) of n is governed by the master equation

$$\overline{n(u, x, t + \tau)} = \overline{\rho(x', t)} p(u|U(x', t)), \tag{6.46}$$

where $x' = x - v_0 u \tau$ and $p(u|U)$ denotes the conditional probability of finding a particle with a velocity u when the local velocity field U is given. From eqn (6.12) we have

$$p(u|U) = P_\xi(u - G(\langle U \rangle)), \tag{6.47}$$

with P_ξ being the probability density of the noise ξ.

Since n is finite, the actual occupation numbers in a given system differ from $\overline{n}$. This fact can be accounted by adding a noise term $\nu \equiv n - \overline{n}$ to the master equation (6.46) as

$$n(u, x, t + \tau) = \rho(x', t) p(u|U(x', t)) + \nu(u, x', t). \tag{6.48}$$

By definition,

$$\overline{\nu} = 0. \tag{6.49}$$

Due to the conservation of the particles

$$\int v \, du = 0 \tag{6.50}$$

holds. Since the assignment of the velocities at each time-step is an *independent* stochastic event for each particle, the actual values of n satisfy binomial statistics: the probability that we have $n(u, x', t)$ particles moving with velocity u out of $\rho(x', t)$ particles is given by

$$\binom{\rho}{n} [p(u|U)]^n [1 - p(u|U)]^{\rho-n}. \tag{6.51}$$

For large enough ρ eqn (6.51) can be well approximated by a Poisson distribution with parameter $\bar{n} = \rho p(u|U)$. Therefore, the distribution function of v depends on ρ, u and U as

$$P_v(x) = \frac{\bar{n}^{x+\bar{n}} e^{-\bar{n}}}{\Gamma(x + \bar{n} + 1)}. \tag{6.52}$$

Thus, from eqn (6.52) we have

$$\overline{v^2} = \bar{n}. \tag{6.53}$$

To derive equations for the time development of the coarse-grained density and velocity fields, we integrate eqn (6.48) according to du and take the Taylor expansion of $n(u, x - v_0 u\tau, t)$ up to second order in x. Using eqns (6.44), (6.47) and integrating eqn (6.48) we obtain

$$\rho(x, t + \tau) = \int du \, \rho(x - v_0 \tau u, t) P_\xi(u - g(x - v_0 \tau u, t))$$

$$+ \int du \, v(u, x - v_0 \tau u, t), \tag{6.54}$$

with $g(x', t) = G(\langle U(x', t) \rangle)$. To evaluate the first integral at the right-hand-side of (6.54), we expand ρ and g as

$$\rho(x - v_0 \tau u) = \rho(x) - v_0 \tau u (\partial_x \rho)(x) + \frac{v_0^2 \tau^2 u^2}{2} (\partial_x^2 \rho)(x) + \cdots \tag{6.55}$$

$$P_\xi(u - g(x - v_0 \tau u)) = P_\xi(u - g(x)) + v_0 \tau u \frac{dP_\xi}{du}(u - g(x))(\partial_x g)(x)$$

$$- \frac{v_0^2 \tau^2 u^2}{2} \left[\frac{d^2 P_\xi}{du^2}(u - g(x))(\partial_x g)^2(x) - \frac{dP_\xi}{du}(u - g(x))(\partial_x^2 g)(x) \right] + \cdots \tag{6.56}$$

where we have omitted for simplicity the time from the notation. Substituting eqns (6.55) and (6.56) into eqn (6.54), we arrive at

$$\int du \rho(x - v_0 \tau u) P_\xi(u - g(x - v_0 \tau u))$$

$$= \rho(x) - v_0 \tau \partial_x (\rho g)(x)$$

$$+ \frac{v_0^2 \tau^2 \sigma^2}{2} (\partial_x^2 \rho)(x) + v_0^2 \tau^2 \left[\rho g (\partial_x^2 g) - \rho (\partial_x g)^2 - 2(\partial_x \rho) g (\partial_x g) \right](x) \tag{6.57}$$

$$+ \cdots \tag{6.58}$$

If $\sigma^2 \equiv \int P_\xi(u)u^2\,du \gg 1$, the eqn (6.57) term, and for $v_0\tau \ll 1$ also the higher-order terms, can be neglected in eqn (6.58).

The expected value of the second integral on the right-hand-side of eqn (6.54) is zero due to eqn (6.49). Its typical magnitude can be estimated by calculating its standard deviation, using eqn (6.53), as

$$\int du_1 \int du_2\, \overline{v(u_1, x - v_0\tau u_1)v(u_2, x - v_0\tau u_2)}$$
$$= \int du_1\, n(u_1, x - v_0\tau u_1) \approx \rho(x), \qquad (6.59)$$

i.e., the typical value of the integral is of the order of $\sqrt{\rho}$, which is small compared to the first term for $\rho \gg 1$.

Thus, in the limit of $v_0\tau \ll 1$, $\sigma^2 \equiv \int P_\xi(u)u^2du \gg 1$ and $\rho \gg 1$, eqn (6.54) simplifies to

$$\partial_t \rho = -v_0\partial_x(\rho U) + D\partial_x^2\rho, \qquad (6.60)$$

with

$$D = v_0^2\tau\sigma^2/2. \qquad (6.61)$$

This equation corresponds to the conservation of the particles, and we have established the connection between its 'macroscopical' parameters v_0, D and the parameters of the underlying microscopic dynamics. Note, that the appearance of the diffusion term is a consequence of the nonvanishing correlation time τ.

In a similar fashion, integrating eqn (6.48) according to $u\,du$, expanding $\langle U \rangle$ as

$$\langle U \rangle(x) = \frac{\int_{x-\Delta}^{x+\Delta} U(x')\rho(x')\,dx'}{\int_{x-\Delta}^{x+\Delta}\rho(x')\,dx'} = U + \frac{\Delta^2}{6}\left[\partial_x^2 U + 2\frac{(\partial_x U)(\partial_x \rho)}{\rho}\right] + \ldots, \qquad (6.62)$$

and using eqn (6.60), one arrives at

$$\partial_t U = f(U) + \mu^2\partial_x^2 U + \alpha\frac{(\partial_x U)(\partial_x \rho)}{\rho} + \zeta, \qquad (6.63)$$

where $f(U) = (G(U) - U)/\tau$, $\mu^2 = (dG/dU)/(6\tau)$, $\alpha = 2\mu^2$ and

$$\zeta = \frac{1}{\rho\tau}\int vu\,du. \qquad (6.64)$$

Note that $f(U)$ is an antisymmetric function with $f(U) > 0$ for $0 < U < 1$ and $f(U) < 0$ for $U > 1$. The noise ζ is *not* negligible in this case, and satisfies

$$\overline{\zeta} = 0 \qquad (6.65)$$
$$\overline{\zeta^2} = \sigma^2/\rho\tau^2. \qquad (6.66)$$

At this point we consider eqns (6.60) and (6.63) with the coefficients μ, α, σ, v_0 and D as the continuum theory describing a large class of SPP models. The nonlinear coupling term $(\partial_x U)(\partial_x \rho)/\rho$ is *specific* to such SPP systems: it is responsible for the slowing down (and eventually the 'turning back') of the particles under the influence of a larger number of particles moving in the opposite direction. When two groups of particles move in the opposite direction, the density increases locally and the velocity decreases at the point at which they meet. Let us consider the particular case where particles move from left to right and the velocity decreases locally while the density increases as x increases (the particles move towards a 'wall' formed between two oppositely moving groups). The term $(\partial_x u)$ is less and the term $(\partial_x \rho)$ is larger than zero in this case. Together they have a negative sign, resulting in the slowing down of the local velocity. This is a consequence of the fact that there are more slower particles (in a given neighbourhood) in the forward direction than faster particles coming from behind, and so the average action experienced by a particle in the point x slows it down.

For $\alpha = 0$ the dynamics of the velocity field U is independent of ρ, and eqn (6.63) is equivalent to the time-dependent Ginsburg–Landau Φ^4 model of spin chains, where domains of opposite magnetization develop at finite temperatures [4]. As numerical simulations demonstrate (Fig. 6.21), for large-enough α and low noise the initial domain structure breaks down and the system becomes organized into a single major group travelling in a spontaneously selected direction.

To study the effect of the nonlinear coupling term $\alpha(\partial_x U)(\partial_x \rho)/\rho$, we now investigate the development of the ordered phase in the deterministic case ($\sigma = 0$) when $c, D \ll 1$ holds. It can be seen that stationary domain-wall solutions exist for any α. In particular, let us denote by ρ^* and U^* the stationary solutions which satisfy $\rho^*(\pm\infty) = 0$, $U^*(\pm\infty) = \mp 1$, $U^*(x) < 0$ for $x > 0$, and $U^*(x) > 0$ for $x < 0$.

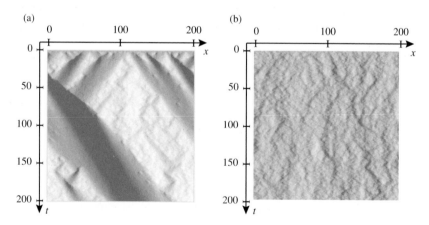

Fig. 6.21 Results of the numerical integration of eqns (6.60) and (6.63). In this space–time (horizontal–vertical) plot the regions of higher densities are indicated by darker shades. For small-enough noise (a) the system is organized into one high-density group travelling in a spontaneously selected direction. In contrast, for higher noise the system remains homogeneous at large length scales (b). (After [27].)

These functions are determined as

$$\ln \frac{\rho^*(x)}{\rho^*(0)} = \frac{v_0}{D} \int_0^x U^*(x') \, dx' \tag{6.67}$$

and

$$\mu^2 \partial_x^2 U^* = -f(U^*) - \alpha \frac{v_0}{D} U^* \partial_x U^*. \tag{6.68}$$

Although stationary solutions exist for any α, they are not always stable against long wavelength perturbations, as the following linear stability analysis shows.

Let us assume that at $t = 0$ we superimpose a $u(x, t = 0)$ perturbation over the $U^*(x)$ stationary solution. Since $v_0, D \ll 1$, the dynamics of ρ is slow, thus $\rho(x, t) = \rho^*(x)$ is assumed. The stationary solutions are metastable in the sense that small perturbations can transform them into equivalent, shifted solutions. Here, by linear stability or instability we mean the existence or inexistence of a stationary solution to which the system converges during the response to a small perturbation. To handle the set of possible metastable solutions we write the perturbation u in the form of $\tilde{u}$ as

$$U(x, t) = U^*(x) + u(x, t) = U^*(x - \xi(t)) + \tilde{u}(x, t), \tag{6.69}$$

i.e.,

$$\tilde{u}(x, t) = u(x, t) + \xi(t)a(x), \tag{6.70}$$

where $a \leq 0$ denotes $\partial_x U^*$ and the position of the domain wall, $\xi(t)$, is defined by the implicit equation

$$U(\xi(t), t) = 0. \tag{6.71}$$

As $U^*(0) = 0$, from eqn (6.71) we have $\tilde{u}(\xi, t) = 0$. The usage of ξ and $\tilde{u}$ is convenient, since the stability of the stationary solution U^* is equivalent to the convergence of $\xi(t)$ as $t \to \infty$.

Substituting eqn (6.69) into eqn (6.60) and taking into account the stationarity of U^* we get

$$-a(x - \xi)\dot{\xi} + \partial_t \tilde{u}(x) = (f' \circ U^*)(x - \xi)\tilde{u}(x) + \mu^2 \partial_x^2 \tilde{u}(x) + \alpha \xi h(x - \xi), \tag{6.72}$$

with $h = a \partial_x^2 \ln \rho^* \geq 0$. To simplify eqn (6.72) let us write $(f' \circ U^*)(x)$ in the form of $g(x) - g_\infty$, where $g_\infty = -\lim_{x \to \pm\infty} f'(U^*(x)) = -f'(\pm 1)$. Furthermore, a moving frame of reference $y = x - \xi$ and a new variable $v(y) = u(x)$ can be defined. With these new notations eqn (6.72) becomes

$$-a(y)\dot{\xi} + \partial_t v(y) = g(y)v(y) - g_\infty v(y) + \mu^2 \partial_x^2 v(y) + \alpha \xi h(y). \tag{6.73}$$

The time development of ξ is determined by the condition $u(\xi) = v(0) = 0$, yielding

$$-a(0)\dot{\xi} = \mu^2 \partial_x^2 v(0) + \alpha \xi h(0). \tag{6.74}$$

Since $v(0) = 0$ and $g > 0$ only in the vicinity of zero, one can show [27] that the gv term has a negligible effect on the solutions of eqns (6.73) and (6.74).

By the Fourier-transformation of eqn (6.73) for the time derivatives of $\xi(t)$ and the nth Fourier moments $\hat{v}_n(t) \sim \partial_x^n v(0, t)$ one can obtain

$$
\frac{d}{dt}
\begin{pmatrix}
-\hat{a}_0 & & & \\
-\hat{a}_2 & 1 & & \\
-\hat{a}_4 & & 1 & \\
\vdots & & & \ddots
\end{pmatrix}
\begin{pmatrix}
\xi \\
\hat{v}_2 \\
\hat{v}_4 \\
\vdots
\end{pmatrix}
=
\begin{pmatrix}
\alpha\hat{h}_0 & -\mu^2 & & \\
\alpha\hat{h}_2 & -g_\infty & -\mu^2 & \\
\alpha\hat{h}_4 & & -g_\infty & -\mu^2 \\
\vdots & & & \ddots
\end{pmatrix}
\begin{pmatrix}
\xi \\
\hat{v}_2 \\
\hat{v}_4 \\
\vdots
\end{pmatrix}.
\tag{6.75}
$$

Expression (6.75) can be further simplified using the relations $\hat{a}_m \ll \hat{a}_n$ and $\hat{h}_m \ll \hat{h}_n$ for $m > n$. For the λ growth rate of the original u perturbation we obtain (see [27] for details)

$$
\lambda_0 = \hat{a}_0 g_\infty < 0,
\tag{6.76}
$$

where λ_+, λ_- satisfy

$$
\lambda_\pm^2 - b\lambda_\pm + q = 0,
\tag{6.77}
$$

where

$$
b = \alpha\hat{h}_0 - \mu^2\hat{a}_2 + g_\infty\hat{a}_0,
\tag{6.78}
$$

$$
q = \alpha\hat{a}_0(\hat{h}_0 g_\infty - \hat{h}_2\mu^2).
\tag{6.79}
$$

If $\alpha = 0$, then $\lambda_+ = 0$ and $\lambda_- = g_\infty\hat{a}_0 - \mu^2\hat{a}_2 < 0$. However, for certain $\alpha > 0$ values $\lambda_+ > 0$ can hold obtained as either $q < 0$ or $b > 0$, i.e.,

$$
\alpha > \alpha_{c,1} = D\frac{3g_\infty - 2g(0)}{2v_0 a(0)}
\tag{6.80}
$$

or

$$
\alpha > \alpha_{c,2} = D\frac{2g_\infty - g(0)}{a(0)(v_0 - D)},
\tag{6.81}
$$

respectively. Thus for $\alpha > \alpha_c = min(\alpha_{c,1}, \alpha_{c,2})$ the stability of the domain-wall solution disappears.

The instability of the domain-wall solutions gives rise to the ordering of the system, as the following simplified picture shows. A small domain of (left-moving) particles moving in the opposite direction to the surrounding (right-moving) ones is *bound to interact* with more and more right-moving particles and, as a result, the domain wall assumes a specific structure which is characterized by a buildup of the right-moving particles on its left side, while no more than the originally present left-moving particles remain on the right side of the domain wall. This process 'transforms' nonlocal information (the size of the corresponding domains) into a *local asymmetry of the particle density* which, through the instability of the walls, results in a leftward motion of the domain wall, and consequently, eliminates the smaller domain.

This can be demonstrated schematically as

$$>>>>>>>>>>>><<<<<<<<<>>>>>>>>>>>>>>$$
$$\qquad\qquad A \qquad\qquad\qquad\qquad B$$

where by $>$ ($<$) we denoted the right-(left-) moving particles. In contrast to the Ising model the A and B walls are very different and have nonequivalent properties. In this situation the B wall will break into a B_1 and B_2, moving in opposite directions, B_1 moving to the left and B_2 moving to the right, leaving the area $B_1 - B_2$ behind, which is depleted of particles.

$$>>>>>>>>>>>><<<<<<\qquad\qquad>>>>>>>>>>$$
$$\qquad A \qquad\qquad\quad B_1 \qquad B_2$$

At the A boundary the two types of particle slow down, while, due to the instability we showed to be present in the system, the wall itself moves in a certain direction, most probably to the right. Even in the other, extremely rare, case (i.e., when the A wall moves to the left), an elimination of the smaller domain ($A - B_1$) takes place since the velocity of the domain wall A is smaller than the velocity of the particles in the 'bulk' and at B_1, where the local average velocity is the same as the preferred velocity of the particles. Thus, the particles tend to accumulate at the domain wall A, which again, through local interactions, leads to the elimination of the domain $A - B_1$.

It is easy to see that the $U = \pm 1$ solutions are absolutely stable against small perturbations; thus it is plausible to assume that the system converges into those solutions even for finite noise.

6.1.6 Hydrodynamic formulation for 2D

To get an insight into the possible analytical treatment of SPP systems in higher dimensions, we now apply the Navier–Stokes equations for a 'fluid' of SPPs, as described in [28]. The two basic equations governing the dynamics of the *noiseless* 'self-propelled fluid' are the continuity equation

$$\partial_t \rho + \nabla(\rho \boldsymbol{v}) = 0 \tag{6.82}$$

and the equation of motion

$$\partial_t \boldsymbol{v} + (\boldsymbol{v} \cdot \nabla)\boldsymbol{v} = \boldsymbol{F}(\boldsymbol{v}) - \frac{1}{\tau_1}\boldsymbol{v} + \frac{1}{\tau_2}(\langle \boldsymbol{v} \rangle - \boldsymbol{v}) - \frac{1}{\rho^*}\nabla p + \nu \nabla^2 \boldsymbol{v}, \tag{6.83}$$

where ρ^* is the effective density of the particles, p is the pressure, ν is the kinematic viscosity, $\boldsymbol{F}(\boldsymbol{v})$ is the *intrinsic* driving force of biological origin and τ_1, τ_2 are time scales associated with velocity relaxation resulting from interaction with the environment and the surrounding SPPs, respectively. Let us now go through the terms of eqn (6.83).

The self-propulsion can be taken into account as a constant-magnitude force acting in the direction of the velocity of the particles as

$$F = \frac{v_0}{\tau_1} \frac{v}{|v|}, \tag{6.84}$$

where v_0 is the speed determined by the balance of the propulsion and friction forces, i.e., v_0 would be the speed of a homogeneous fluid.

Performing a Taylor expansion of the expression $\frac{1}{\tau_2}(\langle v \rangle - v)$ describing local velocity alignment yields

$$\langle v \rangle_\epsilon - v = \frac{\int_{|\xi|<\epsilon} d\xi \left(v\rho + (\xi\nabla)v\rho + \frac{1}{2}(\xi\nabla)^2 v\rho + \cdots \right)}{\int_{|\xi|<\epsilon} d\xi \left(\rho + (\xi\nabla)\rho + \frac{1}{2}(\xi\nabla)^2 \rho + \cdots \right)} - v$$

$$= \frac{\epsilon^2}{6} \left(\frac{\nabla^2(v\rho)}{\rho} - v\frac{\nabla^2\rho}{\rho} \right) + \cdots$$

$$= \frac{\epsilon^2}{6} \left(\nabla^2 v + 2(\nabla v)\frac{\nabla\rho}{\rho} \right) + \cdots, \tag{6.85}$$

which is similar to the 1D case, eqn (6.62). In the following we will consider cases where the density fluctuations are forced to be small. Hence the velocity–density coupling term in eqn (6.85) is negligible, which is a major difference compared to the previously studied 1D system. Thus, if the relative density changes are small, the velocity alignment can be incorporated with an effective viscosity coefficient v^* into the viscous term of eqn (6.83).

In the simplest cases the pressure can be composed of an effective 'hydrostatic' pressure, and an externally applied pressure as

$$p = g^*\rho + p_{\text{ext}}, \tag{6.86}$$

where g^* is a parameter related to the compressibility of the fluid.

Combining eqns (6.82), (6.83), (6.84) and (6.86) one gets the following final form for the equations of the SPP flow:

$$\partial_t \rho + \nabla(\rho v) = 0, \tag{6.87}$$

and

$$\partial_t v + (v \cdot \nabla)v = \frac{v_0}{\tau_1}\frac{v}{|v|} - \frac{1}{\tau_1}v + v^*\nabla^2 v - g^*\nabla\rho + -\frac{1}{\rho^*}\nabla p_{\text{ext}}. \tag{6.88}$$

It will be useful in the following to introduce the characteristic length

$$\lambda = \sqrt{v^*\tau_1} \tag{6.89}$$

and characteristic time

$$T = \frac{\lambda}{v_0} \qquad (6.90)$$

and write eqns (6.87) and (6.88) in the dimensionless form of

$$\partial_{t'}\rho + \nabla'(\rho \boldsymbol{v}') = 0, \qquad (6.91)$$

and

$$\frac{\tau_1}{T}\left(\partial_{t'}\boldsymbol{v}' + (\boldsymbol{v}' \cdot \nabla')\boldsymbol{v}'\right) = \frac{\boldsymbol{v}'}{|\boldsymbol{v}'|} - \boldsymbol{v}' + \nabla'^2\boldsymbol{v}' - g_1'\nabla'\rho + g_2'\nabla'p_{\text{ext}}, \qquad (6.92)$$

where $t' = t/T$, $x' = x/\lambda$, $v' = v/v_0$ and the ∇' operator differentiates with respect to $\boldsymbol{r}' = \boldsymbol{r}/\lambda$. We also introduced the notations $g_1' = T\tau_1 g^*/\lambda^2$ and $g_2' = g_1'/g^*$. In the following we will drop the prime for simplicity.

For certain simple geometries it is possible to obtain analytical stationary solutions of the above equations, supposing incompressibility ($\rho = \text{const}$). Taking $\partial_t \equiv 0$ in eqns (6.91) and (6.92), the following dimensionless equations are obtained:

$$\nabla \boldsymbol{v} = 0 \qquad (6.93)$$

and

$$\frac{\tau_1}{T}(\boldsymbol{v}\nabla)\boldsymbol{v} = \frac{\boldsymbol{v}}{|\boldsymbol{v}|} - \boldsymbol{v} + \nabla^2\boldsymbol{v} - g_2\nabla p_{\text{ext}}. \qquad (6.94)$$

A simple, but relevant boundary condition is realized when the system is confined to a circular area of radius R. Since this is a finite geometry, in the stationary state no net flux is possible, i.e., the flow field must include vortices. In [28] it has been demonstrated that a single vortex is indeed a possible stationary configuration of the system: the velocity field can be expressed in polar coordinates (r, ϕ) as

$$\boldsymbol{v} = v(r)\,\boldsymbol{e}_\phi, \qquad (6.95)$$

where $\boldsymbol{e}_\phi$ is the tangential unit vector. Since $\nabla = \partial_r + (1/r)\partial_\phi$, one can easily see that eqn (6.93) is satisfied for any velocity profile $v(r)$. Substituting eqn (6.95) into eqn (6.94) yields two ordinary differential equations:

$$0 = r^2\frac{d^2v}{dr^2} + r\frac{dv}{dr} - v(1 + r^2) + r^2, \qquad (6.96)$$

and

$$-\frac{1}{r^2}v^2 = -\frac{1}{\rho^*\lambda}\frac{\partial p_{\text{ext}}}{\partial r}. \qquad (6.97)$$

The first equation gives the velocity profile, while the second determines the pressure to be applied to maintain the constant ρ condition.

The boundary conditions for the velocity profile are either $v(0) = 0$ and $v(R) = 0$ (closed boundary at $r = R$) or $v(0) = 0$ and $(dv/dr)|_R = 0$ (free boundary at $r = R$). The solution of eqn (6.96) with the above boundary conditions is given by $I_1(r)$, the modified Bessel function of order one, and $L_1(r)$, the modified Struve function of order one [29]. Thus the velocity profile of a stationary single vortex is

$$v(r) = \alpha I_1(r) - \frac{\pi}{2} L_1(r). \tag{6.98}$$

The parameter α should be chosen to satisfy the boundary condition at $r = R$.

In Fig. 6.22(a) we show velocity profiles for different dimensionless system size R. The maximal velocity decreases for decreasing R (Fig. 6.22(b)). From Fig. 6.22 it is also clear that the minimal size of a vortex is of the order of λ, i.e., one in dimensionless units. Therefore, if $R \gg 1$, then initially many Kosterlitz–Thouless-like vortices are likely to be present in the system.

For further investigations numerical solutions of eqns (6.87) and (6.88) were also calculated in [28] with finite compressibility and $p_{ext} = 0$. The only remaining dimensionless quantity characterizing the system is τ_1/T, which relates the various relaxation times and v_0. The system can be considered overdamped if $\tau_1 \ll T$ holds. Figure 6.23 shows the stationary state for the overdamped equations in the high compressibility ($g_1 = 750$) limit. The length and direction of the arrows show the velocity, while the thickness is proportional to the local density of the fluid. In Fig. 6.23(b) the radial velocity distribution is presented for the vortex shown in Fig. 6.23(a) together with the velocity profile given by eqn (6.98). Rather good agreement is seen; the differences are due to the fact that the numerical system is not perfectly circular. In simulations with $R \gg 1$, multi-vortex states develop (Fig. 6.24), which transform into a single vortex after a long enough time.

Thus, we have demonstrated that – in contrast to the 1D case – the ordering in 2D is not due to the density–velocity coupling term in the expansion of $\langle v \rangle$: the viscosity and the internal driving force terms are sufficient to destabilize the vortices originally present and organize the system into a globally ordered phase. The stability of that

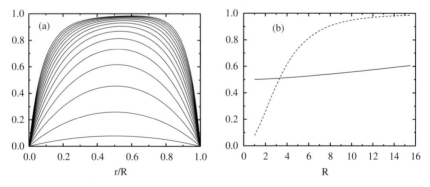

Fig. 6.22 (a) Velocity profiles in a vortex for various values of the dimensionless system size, R (bottom curve: $R = 1$, top curve: $R = 16$). (b) Position (solid line) and value (dashed line) of the maximal velocity for the profiles shown in (a) as a function of R. (After [28].) Copyright 1997 reprinted with permission from Elsevier Science.

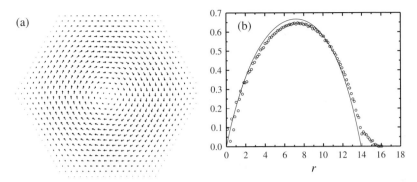

Fig. 6.23 (a) A numerically generated vortex for dimensionless system size $R = 4.4$ and $g_1 = 750$. The length of the arrows is proportional to the local velocity, while their thickness is proportional to the density. (b) The measured (circles) and the theoretical (solid line) velocity profile for the vortex shown in (a) (after [28]). Copyright 1997 reprinted with permission from Elsevier Science.

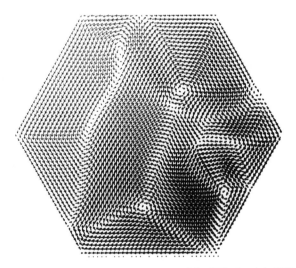

Fig. 6.24 Transient multiple vortex state at $R = 25.3$, $g_1 = 128$ (after [28]). Copyright 1997 reprinted with permission from Elsevier Science.

ordered phase against a finite amount of fluctuations has been shown by Tu and Toner, as we discuss in the next section.

6.1.7 The existence of long-range order

The first theory describing the full nonlinear higher-dimensional dynamics was presented in [11, 30]. Tu and Toner followed the historical precedent [31] of the Navier–Stokes equation by deriving the continuum, long-wavelength description *not* by explicitly coarse graining the microscopic dynamics, as we did it for the one-dimensional system, but, rather, by writing down the most general continuum equations of motion for v and ρ consistent with the symmetries and conservation laws of the problem. This approach allows us to introduce a few phenomenological parameters (like the viscosity in the Navier–Stokes equation), whose numerical values will depend on the detailed microscopic behaviour of the particles. The terms in the equations describing the large-scale behaviour, however, should depend only on symmetries and conservation laws, and *not* on the microscopic rules.

The only symmetry of the system is rotation invariance: since the particles lack a compass, all direction of space are equivalent to other directions. Thus, the 'hydrodynamic' equation of motion cannot have built into it any special direction chosen 'a priori'; all directions must be spontaneously selected. Note that the model does *not* have Galilean invariance: changing the velocities of all the particles by some constant boost v_b does *not* leave the model invariant.

To reduce the complexity of the equations of motion still further, a spatial-temporal gradient expansion can be performed keeping only the lowest-order terms in gradients and time derivatives of v and ρ. This is motivated and justified by the aim of considering *only* the long-distance, long-time properties of the system. The resulting set of equations is

$$\partial_t v + \lambda_1 (v\nabla)v + \lambda_2 (\nabla v)v + \lambda_3 \nabla(|v|^2)$$
$$= \alpha v - \beta |v|^2 v - \nabla P + D_L \nabla(\nabla v) + D_1 \nabla^2 v + D_2 (v\nabla)^2 v + \xi$$

$$\partial_t \rho + \nabla(\rho v) = 0, \tag{6.99}$$

where the $\alpha, \beta > 0$ terms make v have a non-zero magnitude, $D_{L,1,2}$ are diffusion constants and ξ is uncorrelated Gaussian random noise. The λ terms on the left-hand side of eqn (6.99) are the analogues of the usual convective derivative of the coarse-grained velocity field v in the Navier–Stokes equation. Here the absence of Galilean invariance allows all *three* combinations of one spatial gradient and two velocities that transform like vectors; if Galilean invariance *did* hold, it would force $\lambda_2 = \lambda_3 = 0$ and $\lambda_1 = 1$. However, Galilean invariance does *not* hold, and so all three coefficients can be non-zero phenomenological parameters whose values are determined by the microscopic rules. The pressure P depends on the local density only, as given by the expansion

$$P(\rho) = \sum_n \sigma_n (\rho - \rho_0)^n. \tag{6.100}$$

It is possible to treat the whole problem analytically using the dynamical renormalization group to show the existence of an ordered phase in 2D and to extract exponents characterizing the velocity–velocity and density–density correlation functions [30]. The most dramatic result is that an intrinsically nonequilibrium and nonlinear feature, namely, convection, suppresses fluctuations of the velocity v at long wavelengths, making them much smaller than the analogous fluctuations found in ferromagnets, for all spatial dimensions $d < 4$. Specifically, the measure of the fluctuations, $C_C(\boldsymbol{R})$, defined as:

$$C_C(\boldsymbol{R}) = C(\boldsymbol{R}) - \lim_{|\boldsymbol{R}'| \to \infty} C\left(\boldsymbol{R}'\right), \tag{6.101}$$

with

$$C(\boldsymbol{R}) \equiv \langle \boldsymbol{v}(\boldsymbol{R} + \boldsymbol{r}, t) \cdot \boldsymbol{v}(\boldsymbol{r}, t) \rangle \tag{6.102}$$

and

$$C(\boldsymbol{R} \to \infty) = \phi^2, \tag{6.103}$$

decays to zero much more rapidly, as $|\boldsymbol{R}| \to \infty$, than the analogous correlation function in magnets. Quantitatively, for points whose separation $\boldsymbol{R} \equiv \boldsymbol{R}_\perp$ lies perpendicular to the mean direction of motion

$$C_C(\boldsymbol{R}) \propto R_\perp^{2\chi} \tag{6.104}$$

holds, where the universal exponent is

$$\chi = -\tfrac{1}{5} \tag{6.105}$$

exactly, in $d = 2$, and is smaller than its equilibrium value, $1 - d/2$, in magnetic systems for all $d < 4$.

The physical mechanism for this suppression of fluctuations is easy to understand: increased fluctuations in the direction of motion of different parts of the system actually *enhance* the exchange of information between those different parts. This exchange, in turn, suppresses those very fluctuations, since the interactions between particles tend to make them all move in the same direction.

These nonequilibrium effects also lead to a spatial anisotropy of scaling between the direction along ($\parallel$) and those orthogonal to ($\perp$) the mean velocity $\langle \boldsymbol{v} \rangle$. The physical origin of the anisotropy is also simple: if particles make small errors $\delta\theta$ in their direction of motion, their random motion *perpendicular* to the mean direction of motion $\langle \boldsymbol{v} \rangle$ is much larger than that *along* $\langle \boldsymbol{v} \rangle$; the former is proportional to $\delta\theta$, while the later is proportional to $1 - \cos\delta\theta \sim \delta\theta^2$. As a result, *any* equal-time correlation function in the system of *any* combination of fields crosses over from dependence purely on $|\boldsymbol{R}_\perp|$ to dependence purely on $R_\parallel$ when

$$\frac{R_\parallel}{\ell_0} \approx \left(\frac{|\boldsymbol{R}_\perp|}{\ell_0} \right)^\zeta, \tag{6.106}$$

where ℓ_0 is the interaction range. The universal anisotropy exponent is *exactly*

$$\zeta = \tfrac{3}{5}, \tag{6.107}$$

for $d = 2$, and is < 1 for all $d < 4$.

6.2 Correlated motion of pedestrians

D. Helbing and P. Molnár

6.2.1 Introduction

To physicists, nonequilibrium phase transitions are very fascinating phenomena. A great variety of self-organized phenomena is, for example, found in driven granular media. This includes the evolution of density waves or convection patterns and the segregation of grains of different sizes (depending on the respective experimental conditions) [32, 33]. Recently, particular interest has arisen in spatio-temporal, collective patterns of motion which are formed in systems of so-called 'motorized' or 'self-driven' particles [6, 17, 22, 25]. Whereas such systems are exceptional in physical systems (like discs moving on an air table), they are frequently found in social or biological systems. Thus, it is not surprising that many examples of spontaneous structure formation can be found there. It turned out that they can be often understood with methods from statistical physics and nonlinear dynamics, even in a quantitative way. Examples are the movement of fish or bird swarms, pattern formation in colonies of bacteria or slime molds, or the development of ant trails [11, 34, 35].

A variety of self-organization phenomena is also found in social systems [36]. However, a serious problem in describing these systems originates from the fact that many variables which influence human behaviour are hardly measurable. For this reason pedestrian crowds are an ideal object of social research: all essential quantities like places, speeds and walking directions of pedestrians as well as locations of obstacles and attractions, etc. can be easily and exactly determined. Moreover, a large amount of data has already been collected [37, 38].

6.2.2 Pedestrian dynamics

Pedestrian crowds have been studied for more than three decades now [37–39], e.g., by means of time-lapse films, photographs, and other evaluation methods (§6.2.2). The main goal of these studies was to develop guidelines for planning and designing pedestrian facilities. These usually have the form of regression relations [37] which are, however, not very well suited for the prediction of pedestrian flows in pedestrian precincts or buildings with an exceptional architecture. Therefore, a number of simulation models have been proposed, e.g., *queueing models* [40] and models for the route-choice behaviour of pedestrians [41].

None of these concepts adequately takes into account the self-organization effects occurring in pedestrian crowds. These may, however, lead to unexpected obstructions due to mutual disturbances of pedestrian flows. More promising with regard to this is the approach by Henderson. He conjectured that pedestrian crowds behave in a similar way to gases or fluids [42], which could be partly confirmed (see below). However, a realistic gas-kinetic or fluid-dynamic theory must contain corrections due to the special interactions (i.e., avoidance and deceleration manoeuvres) which, of course, do not conserve momentum and energy. Although such a theory can actually be formulated [43], for practical applications a direct simulation of *individual* pedestrian motion is favourable, since a numerical solution of the fluid-dynamic equations is very difficult. As a consequence, current research focuses on the *microsimulation* of pedestrian crowds. In this connection, a *behavioural-force model* of individual pedestrian dynamics was recently suggested [44–49] (see below), a simple forerunner of which was proposed by Gipps and Marksjö [50]. The model is based on coupled nonlinear Langevin equations, and it is closely related to the *molecular-dynamic models* of granular flows [33].

Observations

We have studied pedestrian motion for several years and evaluated a number of time-lapse video films [51]. Despite the sometimes more or less 'chaotic' appearance of individual pedestrian behaviour, some regularities can be found. The following results mainly apply to purposeful pedestrians who aim for a certain destination. They are not valid for aimless pedestrians, like tourists, who stroll around (showing other rules of motion, cf. [45]). In addition, children move in a more irregular way than that described below, since they still need to learn optimal strategies of motion which are used more or less automatically later.

Our observations can be summarized as follows:

1. Pedestrians normally choose the shortest route to their next destination which has, therefore, the shape of a polygon. If alternative routes have the same length, a pedestrian prefers the one where he/she can go straight ahead as long as possible, and change the direction as late as possible, provided that the alternative route is not more attractive (due to less noise, more light, friendlier environment, less waiting time at traffic lights, etc.). A pedestrian feels a strong aversion to taking detours or moving in the opposite direction to the desired walking direction, even if the direct way is crowded.

2. Pedestrians prefer to walk with an individual desired speed, which corresponds to the most comfortable walking speed, as long as it is not necessary to move faster in order to reach the destination in time. The desired speeds within pedestrian crowds are Gaussian distributed with a mean value of about 1.34 m/s and a standard deviation of 0.26 m/s [52].

3. Pedestrians keep a certain distance away from other pedestrians and borders (of streets, walls, and obstacles). This distance is smaller when a pedestrian hurries, and it decreases with growing pedestrian density. Resting individuals (waiting on a railway platform for a train, sitting in a dining hall, or lying at a beach) are uniformly distributed over the available area if there are no acquaintances among the individuals. Pedestrian density increases (i.e., interpersonal distances lessen) around particularly attractive places. It decreases with growing velocity variance

(e.g., on a dance floor, cf. [43]). Individuals knowing each other may form groups which are entities that behave in a similar way to single pedestrians. Group sizes are Poisson distributed [53].

4. Pedestrians normally do not reflect their behavioural strategy in every situation anew but act more or less automatically (as an experienced car driver does). This becomes obvious when pedestrians cause delays or obstructions, e.g., by entering an elevator or underground even though others may still be trying to get off.

Additionally, we found that, at medium and high pedestrian densities, the motion of pedestrian crowds shows some striking analogies with the motion of gases and fluids [51]:

- Footsteps of pedestrians in snow appear to be related to streamlines of fluids.
- At borderlines between opposite directions of walking one can observe viscous fingering.
- When standing pedestrian crowds need to be crossed, the moving pedestrians form river-like streams (Fig. 6.25).
- The propagation of shock waves can be found in dense pedestrian crowds which push forward.
- Sometimes, the existence of pedestrian-free bubbles is observed.

Apart from these phenomena, there are some analogies with granular flows:

- The velocity profile is flat, and the Hagen–Poiseuille law does not hold [43].
- As in segregation or stratification phenomena in granular media [54], pedestrians spontaneously organize in lanes of uniform walking direction, if the pedestrian density exceeds a certain critical value (Fig. 6.26).

Fig. 6.25 The long-term photograph of a standing crowd in front of a cinema shows that crossing pedestrians form a river-like stream (from [51]).

Fig. 6.26 At sufficiently high densities, pedestrians form lanes of uniform walking direction (from [44, 67]).

- At bottlenecks (e.g., corridors, staircases, or doors), the passing direction of pedestrians oscillates with a frequency that increases with the width and shortness of the bottleneck. This is not only analogous to the fluid-dynamic 'saline oscillator' but also to the granular 'ticking hour glass' [55].

In the following, we will focus on the explanation of self-organization phenomena like the ones mentioned above.

The behavioural-force concept

Many people have the feeling that human behaviour is 'chaotic' or at least very irregular and not predictable. This is possibly true for behaviours that are found in complex situations. However, at least for relatively simple situations *stochastic* behavioural models may be developed, if we restrict ourselves to the description of behavioural *probabilities* that can be found in a huge population (group) of individuals [36]. This idea has been followed by the *gaskinetic* pedestrian model [43].

Another approach for modelling behavioural changes has been suggested by Lewin [56]. According to his idea, behavioural changes are guided by so-called *social fields* or *social forces*, which has been put into mathematical terms by Helbing [57]. In the following, we will examine how this idea could be applied to the description of pedestrian behaviour.

Figure 6.27 is a schematic representation of the processes that lead to behavioural changes. According to this, a sensory *stimulus* leads to a behavioural *reaction* that depends on the personal aims and is chosen from a set of behavioural alternatives with the objective of utility maximization.

Table 6.1 suggests a classification of stimuli. In complex or new situations, the resulting behaviour varies a lot because of the uncertainty of the information underlying the decision. On the other hand, in simple or standard situations, the behaviour will more or less correspond to the *optimal* one. Consequently, the behaviour is accurately predictable. It may even be more or less *'automatic'* for an experienced driver. If the considered behaviours can be represented by points in a (perhaps, abstract) space in which they change continuously in the course of time, they can be described by equations of motion. The functions describing the temporal change of the points in space could be interpreted as the driving forces of this motion. In the following, they will be called *behavioural forces,* since they reflect behavioural changes in response to the respective environment. Previous publications used the term *social force,* since behavioural forces often correspond to social interactions.

Note that behavioural forces differ significantly from forces in physics. In contrast to the latter, they do not fulfil Newton's law *actio = reactio*. Moreover, behavioural forces are not exerted by the environment, but the individuals behave *as if* they would feel them. However, they are produced by the individuals themselves as a result of complex mental, psychological, and physical processes (Fig. 6.27). Behavioural forces can be also imagined as quantities that describe concrete *motivations to act* (including the strength and direction of the intended behavioural changes). In the

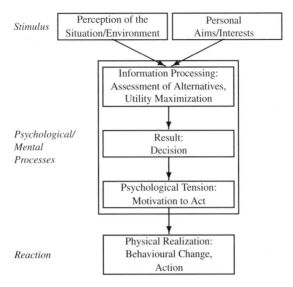

Fig. 6.27 Schematic representation of processes leading to behavioural changes [46]. The human freedom of decision-making is normally not used in an arbitrary or random way. In many situations human behaviour shows rather a certain amount of regularity, which is related to its usefulness and can often be understood as result of utility maximization.

Table 6.1
Classification of behaviours according to their complexity [46].

Stimulus	Simple/Standard Situations	Complex/New Situations
Reaction	Automatic Re-action, 'Reflex'	Result of Evaluation, Decision Process
Characterization	Well Predictable	Probabilistic
Modelling Concept	Social Force Model, etc.	Decision Theoretical Model, etc.
Example	Pedestrian Motion	Destination Choice by Pedestrians

case of pedestrian behaviour, they correspond to real acceleration or deceleration forces as reactions to the perceived information that pedestrians obtain about their environment.

The behavioural-force model of pedestrian motion

Pedestrians are *used* to the situations they are normally confronted with. For this reason, their behavioural strategies are determined by their experience of which reaction to a certain stimulus (situation) will be the *best*. Consequently, their reactions are usually rather 'automatic' and accurately predictable. In the following, we will describe these by a specific behavioural force $f_\alpha(t)$, which represents the different influences (of the environment and other pedestrians) on the behaviour of a pedestrian α. It determines the temporal change $d\boldsymbol{v}_\alpha/dt$ of his/her *actual velocity* $\boldsymbol{v}_\alpha = d\boldsymbol{r}_\alpha/dt$ together with a *fluctuation term* $\boldsymbol{\xi}_\alpha(t)$, which delineates random behavioural variations (arising from accidental or deliberate deviations from the usual rules of motion):

$$\frac{d\boldsymbol{v}_\alpha}{dt} = f_\alpha(t) + \boldsymbol{\xi}_\alpha(t) \tag{6.108}$$

The behavioural force $f_\alpha(t)$ is the sum of several force terms which correspond to the different influences simultaneously affecting the behaviour of pedestrian α:

$$f_\alpha(t) = f_\alpha^0(\boldsymbol{v}_\alpha) + f_{\alpha B}(\boldsymbol{r}_\alpha) + \sum_{\beta(\neq\alpha)} f_{\alpha\beta}(\boldsymbol{r}_\alpha, \boldsymbol{v}_\alpha, \boldsymbol{r}_\beta, \boldsymbol{v}_\beta) + \sum_i f_{\alpha i}(\boldsymbol{r}_\alpha, \boldsymbol{r}_i, t). \tag{6.109}$$

The different contributions are motivated as follows:

- Each pedestrian wants to walk with an individual *desired speed* v_α^0 in the direction $\boldsymbol{e}_\alpha$ of his/her next destination. Deviations of the *actual velocity* $\boldsymbol{v}_\alpha$ from the *desired velocity* $\boldsymbol{v}_\alpha^0 = v_\alpha^0 \boldsymbol{e}_\alpha$ due to disturbances (by obstacles or avoidance manoeuvres) are corrected within the so-called '*relaxation time*' τ_α:

$$f_\alpha^0(\boldsymbol{v}_\alpha) = \frac{1}{\tau_\alpha}\left(v_\alpha^0 \boldsymbol{e}_\alpha - \boldsymbol{v}_\alpha\right). \tag{6.110}$$

In order to compensate for delays, the desired speed $v_\alpha^0(t)$ is often increased in the course of time, since it is given by the remaining distance, divided by the remaining time, until the respective destination is reached. (For example, if someone recognizes that he/she has forgotten the theatre tickets, so that the remaining distance suddenly increases by the additional way home and back to the present place, the desired velocity and actual walking speed will increase discontinuously!)

- Pedestrians keep some distance from borders in order to avoid the risk of getting hurt. The closer the border is, the more uncomfortable a pedestrian feels. This effect can be described by a repulsive force $f_{\alpha B}$ which decreases monotonically with the distance $\|\boldsymbol{r}_\alpha - \boldsymbol{r}_B^\alpha\|$ between the position $\boldsymbol{r}_\alpha(t)$ of pedestrian α and the nearest point $\boldsymbol{r}_B^\alpha$ of the border. In the simplest case this force can be expressed in terms of a repulsive potential V_B:

$$f_{\alpha B}(\boldsymbol{r}_\alpha) = -\nabla_{\boldsymbol{r}_\alpha} V_B(\|\boldsymbol{r}_\alpha - \boldsymbol{r}_B^\alpha\|). \tag{6.111}$$

Similar repulsive force terms $f_{\alpha\beta}(\boldsymbol{r}_\alpha, \boldsymbol{v}_\alpha, \boldsymbol{r}_\beta, \boldsymbol{v}_\beta)$ can describe how each pedestrian α keeps a situation-dependent distance from the other pedestrians β. This reflects the tendency to respect a *private sphere* ('*territorial effect*') and helps to avoid collisions in cases of sudden velocity changes. Note that these repulsive forces are not symmetric, since pedestrians hardly react to the situation behind them.

Moreover, they are strongly anisotropic, since pedestrians need more space in the walking direction than perpendicular to it. Usually, the space requirements for the next step are taken into account. As a consequence, the repulsive interaction forces are also velocity-dependent.

- Pedestrians show a certain joining behaviour. For example, families, friends or tourist parties often move in groups. In addition, pedestrians are sometimes attracted by window displays, sights, special performances (street artists) or unusual events at positions r_i. Both situations can be modelled by time-dependent attractive forces $f_{\alpha i}(r_\alpha, r_i, t)$ in a similar way to repulsive effects, but with an opposite sign and a longer range of interaction.

For a more detailed discussion and a concrete mathematical specification of the force terms see [44, 46–49]. However, we found that the collective phenomena occurring in pedestrian crowds (see §6.2.2) are not very sensitive to the choice of the concrete force model or certain parameters [58, 59].

Equilibria between different forces

Some of the observations discussed above can be easily explained by looking at the equilibria between certain behavioural forces, for which the acceleration dv_α/dt vanishes. For example, focusing on pedestrians waiting on a railway platform, sitting in a restaurant, or lying on a beach, the individual velocities v_α are all zero. Therefore, in the absence of special attractions, the positions r_α of the pedestrians follow from

$$\sum_{\beta(\neq\alpha)} f_{\alpha\beta}(r_\alpha, 0, r_\beta, 0) = 0. \tag{6.112}$$

Since the repulsive forces of different pedestrians are usually comparable, we get more or less equal distances between them, in agreement with the empirically observed uniform distribution. Therefore, the distances between stationary pedestrians are mainly determined by the available area which is to be shared by them.

In the case of additional attraction effects $f_{\alpha i}$ (like the stage in a rock concert), eqn (6.112) must be supplemented by the corresponding attraction forces:

$$\sum_{\beta(\neq\alpha)} f_{\alpha\beta}(r_\alpha, 0, r_\beta, 0) + \sum_i f_{\alpha i}(r_\alpha, r_i) = 0. \tag{6.113}$$

In accordance with the observations, this clearly leads to more crowded regions close to the attractions at positions r_i.

Focusing on situations where a pedestrian α cannot overtake a slower pedestrian β with velocity v_β, we have to set up the equation for the equilibrium between the acceleration force f_α^0 and the repulsive force $f_{\alpha\beta}$. Because of $v_\alpha = v_\beta$, this yields

$$\frac{1}{\tau_\alpha}\left(v_\alpha^0 e_\alpha - v_\beta\right) + f_{\alpha\beta}(r_\alpha, v_\beta, r_\beta, v_\beta) = 0, \tag{6.114}$$

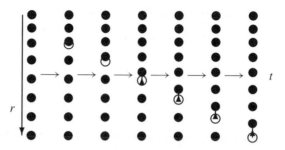

Fig. 6.28 When the front of a queue (top) is stopped, one can often observe the following phenomenon (from [44, 60]). After some time, one of the waiting pedestrians begins to move forward a little, since his/her desired velocity grows, leading to a smaller equilibrium distance to the pedestrian in front. This causes the successor to follow up (and so on), leading to a wave-like propagation of the gap to the end of the queue and to a compaction of the queue (which decreases the comfort of waiting!). Thereby, the tendencies of all individuals to move forward a little (in accordance with their steadily decreasing preferred equilibrium distance) add up towards the end of the queue, giving rise to larger and larger following-up distances. (A more detailed explanation of this phenomenon, considering the minimal length of pedestrian strides, is presented in [45].)

where the actual velocity v_β, the distance vector $(r_\beta - r_\alpha)$, and the repulsive force $f_{\alpha\beta}$ are parallel to the desired direction e_α of walking. We therefore have

$$\frac{v_\alpha^0 - v_\beta}{\tau_\alpha} + f_{\alpha\beta}(r_\alpha, v_\beta e_\alpha, (r_\alpha + \Delta r_{\alpha\beta} e_\alpha), v_\beta e_\alpha) = 0. \qquad (6.115)$$

From this it becomes obvious that pedestrians keep a shorter distance $\Delta r_{\alpha\beta} = \|r_\beta - r_\alpha\|$ apart, when the difference between their own desired velocity v_α^0 and the speed v_β of the preceding pedestrian is larger. This corresponds to the well-known pushing behaviour of pedestrians.

Combined with the growth of the desired speed due to delays, an interesting phenomenon is observed in pedestrian queues [45], as illustrated by Figure 6.28.

By looking at the equilibrium between the acceleration force f_α^0 and an attractive force $f_{\alpha i}$, we can calculate how long a pedestrian joins an attraction at position r_i:

$$\frac{v_\alpha^0(t)e_\alpha - 0}{\tau_\alpha} + f_{\alpha i}(r_\alpha, r_i, t) = 0. \qquad (6.116)$$

In the course of time, the desired speed $v_\alpha^0(t)$ will rise, whereas the interest in the attraction (i.e., the magnitude of $f_{\alpha i}$) tends to decrease. So, at a certain moment $t = t_{\alpha i}$, the attractiveness $f_{\alpha i}$ of position r_i will be compensated for by the tendency f_α^0 to get ahead.

Equilibrium considerations are also useful for appropriately specifying the model parameters, since certain plausibility criteria must be met. For example, pedestrians normally should not move in the opposite direction to their desired walking directions. This implies

$$\frac{v_\alpha^0}{\tau_\alpha} \approx \frac{v_\beta^0}{\tau_\beta}. \qquad (6.117)$$

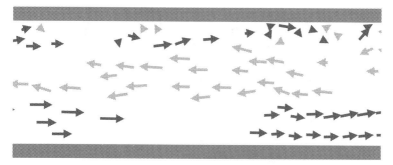

Fig. 6.29 In crowds of oppositely moving pedestrians, one can observe the formation of varying lanes consisting of pedestrians with the same desired direction of motion (from [44, 47–49]; see Java applet at `http://www.helbing.org/pedestrians/corridor.html` and Fig. 6.26). This is even the case if interacting pedestrians avoid each other with the same probability on the right-hand side and on the left-hand side. The reason for lane formation is the related decrease in the frequency of necessary deceleration and avoidance manoeuvres, which increases the efficiency of the pedestrian flow. (The positions, directions, and lengths of the arrows represent the places, walking directions and speeds of pedestrians.)

Simulation of pedestrian crowds and self-organization

The behavioural-force model of pedestrian dynamics has been simulated on a computer for a large number of interacting pedestrians confronted with different situations. Despite the fact that the proposed model is very simple, it describes a lot of observed phenomena very realistically. In particular, under certain conditions the *self-organization* of collective behavioural patterns can be observed. 'Self-organization' means that these patterns are not externally planned, prescribed or organized, e.g., by traffic signs, laws or behavioural conventions. Instead, the spatio-temporal patterns emerge due to the nonlinear interactions of pedestrians. Our model (according to which individuals behave rather automatically) can explain the self-organized patterns described in the following without assuming strategic considerations, communication, or imitative behaviour of pedestrians. All these collective patterns of motion are symmetry-breaking phenomena, although the model was formulated completely symmetrically with respect to the right-hand and the left-hand side.

1. Above a critical pedestrian density our simulations reproduce the empirically observed formation of lanes consisting of pedestrians with the same desired walking direction [44–49] (cf. Fig. 6.29). These lanes are dynamically varying. Their number depends on the width of the street and on pedestrian density.

    The conventional interpretation of lane formation assumes that pedestrians tend to walk on the side which is prescribed in vehicular traffic. However, our model can explain lane formation even without assuming a preference for *any* side. The mechanism of lane formation can be understood as follows [59]. Pedestrians moving against the stream or in areas of mixed directions of motion will have frequent and strong interactions. In each interaction, the encountering pedestrians move

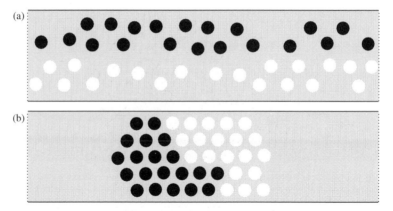

Fig. 6.30 Simulation of 20 pedestrians moving from left to right (black) who interact with 20 pedestrians moving from right to left (white) in a circular street at different noise intensities. (a) Lanes of uniform directions of motion forming for small noise intensity. (b) Blocked, crystallized state resulting for large noise intensity. (From [59].)

a little aside in order to pass each other. This sidewards movement tends to separate oppositely moving pedestrians. Moreover, pedestrians moving in uniform lanes will have very rare and weak interactions. Hence, the tendency to break up existing lanes is negligible when the fluctuations are small. Furthermore, the most stable configuration corresponds to a state with a minimal interaction rate and is related to a maximum efficiency of motion [58] (see eqn (6.118)).

2. We have also investigated the role of fluctuations on pedestrian dynamics. While for small noise amplitudes θ and moderate densities our simulations lead to the formation of lanes (cf. Fig. 6.30(a)), jamming occurs for relatively large numbers N of pedestrians, depending on the respective initial and/or boundary conditions. In this jammed state, the pedestrians were organized in a hexagonal lattice structure, very much as in a crystal (cf. Fig. 6.30(b)). Hence, although the system likes to maximize its efficiency [58], it ends up with minimal efficiency due to noise-induced crystallization. The role of noise here is to destroy the energetically more favourable fluid state and to drive the system 'uphill' towards a jammed state of higher internal energy! Crystallization occurs in this jammed phase, because it yields the densest packing.

We point out that there are pedestrian densities at which we observe *bistability*. That is, we find lanes for small fluctuations, but a crystallized state, if only the noise amplitude (the 'temperature') is sufficiently increased (cf. Fig. 6.30). Therefore, one can drive the system from a 'fluid' state with lanes of uniform directions of motion to a 'frozen' state *just by increasing the fluctuations* (cf. Fig. 6.31). We call this paradoxical transition 'freezing by heating' [59].

We would like to point out that 'freezing by heating' is likely to be relevant to situations involving pedestrians under extreme conditions (panics). Imagine

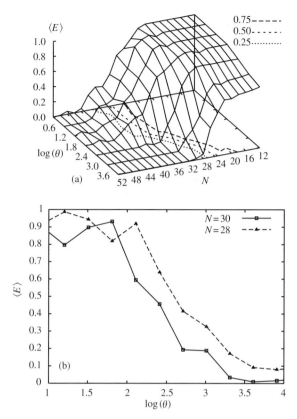

Fig. 6.31 (a), (b): The ensemble-averaged efficiency $\langle E \rangle$ of the system (see eqn (6.118)) as a function of the particle number N and the noise intensity θ on a logarithmic scale. Shown are averages over 25 simulation runs with different random seeds (from [59]). In (a), broken lines are equipotential lines belonging to the labelled levels. The decrease of efficiency from values close to 1 to values around 0 with increasing noise intensity but constant number of particles indicates the transition from the fluid to the crystallized state that we call 'freezing by heating'.

a very smoky situation, caused by a fire, in which people do not know which is the right way to escape. When panicking, people will just try to get ahead, with a reduced tendency to follow a certain direction. Thus, fluctuations will be very large, which can lead to fatal blockings.

3. In our simulations, we also observed oscillatory changes of the walking direction at narrow passages [44–49] (cf. Fig. 6.32). The conventional interpretation for a change of the walking direction is that, after some time, a pedestrian does the kindness of giving precedence to a waiting pedestrian with an opposite walking direction. This, however, cannot explain the increase of oscillation frequency with passage width.

    The mechanism leading to alternating flows is the following. Once a pedestrian is able to pass the narrowing (door, staircase, etc.), pedestrians with the same

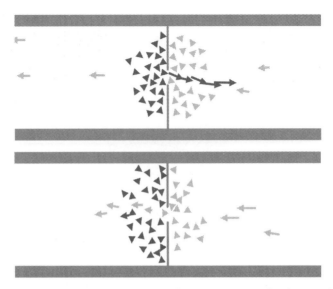

Fig. 6.32 At narrow passages one finds an oscillation of the passing direction (from [44, 47–49], see Java applets at http://www.helbing.org/Pedestrians/Door.html). When a pedestrian is able to pass the door, normally another pedestrian can follow him/her easily (above). However, the pedestrian stream arising in this way will stop after some time due to the pressure from the other side of the passage. A short time period later, a pedestrian will pass the door in the opposite direction, and the process continues as outlined before (see below).

walking direction can easily follow, which is particularly clear for long passages. By this, the number and 'pressure' of waiting and pushing pedestrians becomes less than on the other side of the narrowing, where, consequently, the chance to occupy the passage grows. This leads to a deadlock situation after some time which is followed by a change of the walking direction. Capturing the passage is easier if it is broad and short, so that the walking direction changes more frequently.

4. At intersections our simulations show the temporary emergence of unstable roundabout traffic [44, 47–49] (Fig. 6.33), which is similar to the emergent rotating mode found for self-driven particles [6, 17, 22, 25]. The rotation direction is alternating. Roundabout traffic is connected with small detours but decreases the frequency of necessary deceleration, stopping and avoidance manoeuvres considerably, so that pedestrian motion becomes more efficient on average.

Optimization of pedestrian facilities

The emerging pedestrian flows decisively depend on the geometry of the boundaries. They can already be simulated on a computer in the planning phase of pedestrian facilities. Their configuration and shape can be systematically varied (e.g., by evolutionary algorithms [61]; see Fig. 6.34) and evaluated on the basis of particular mathematical

Fig. 6.33 At intersections one is confronted with various alternating collective patterns of motion which are very short-lived and unstable. For example, phases of temporary roundabout traffic (above, from [44, 47–49]) alternate with phases during which the intersection is crossed in the 'vertical' or 'horizontal' direction. The efficiency of pedestrian flow can be considerably increased by putting an obstacle in the centre of the intersection, since this favours smoothly flowing roundabout traffic compared with competing, inefficient patterns of motion.

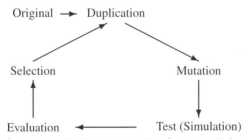

Fig. 6.34 Illustration of evolutionary algorithms. An original configuration is duplicated. Then, the copies are mutated, and their performance is evaluated via simulations. The best copies are selected, mutated, etc., until the configuration has reached an optimal or at least sufficiently good performance. An example of the evolutionary optimization of a bottleneck is presented in Fig. 6.37.

performance measures [49]. For example, the efficiency measure

$$E = \frac{1}{N} \sum_{\alpha} \overline{\frac{\boldsymbol{v}_\alpha \cdot \boldsymbol{e}_\alpha}{v_\alpha^0}} \quad (0 \leq E \leq 1) \tag{6.118}$$

(where N is the number of pedestrians α and the bar denotes a time average) describes the average fraction of desired speed with which pedestrians actually approach their destinations. The measure of discomfort

$$D = \frac{1}{N} \sum_\alpha \frac{\overline{(v_\alpha - \overline{v_\alpha})^2}}{(v_\alpha)^2} = \frac{1}{N} \sum_\alpha \left(1 - \frac{\overline{v_\alpha}^2}{(v_\alpha)^2} \right) \qquad (0 \le D \le 1) \qquad (6.119)$$

reflects the frequency and degree of sudden velocity changes, i.e., the level of discontinuity of walking due to necessary avoidance manoeuvres. Hence, the optimal configuration regarding the pedestrian requirements is the one with the highest values of efficiency and comfort $C = 1 - D$.

During the optimization procedure, some or all of the following can be varied:

(1) the location and form of planned buildings;

(2) the arrangement of walkways, entrances, exits, staircases, elevators, escalators and corridors;

(3) the shape of rooms, corridors, entrances, and exits;

(4) the function and time schedule of room usage.

The proposed optimization procedure can not only be applied to the design of new pedestrian facilities but also to a reduction of existing bottlenecks by suitable modifications. Here, we discuss four examples:

1. At high pedestrian densities, the lanes of uniform walking direction tend to disturb each other: impatient pedestrians try to use any gap for overtaking, which often leads to subsequent obstructions of the opposite walking directions. The lanes can be stabilized by series of trees or columns in the middle of the road (see Fig. 6.35(a)) which, in walking direction, looks similar to a wall (cf. Fig. 6.36). Also, it takes a detour to reach the other side of the permeable wall, which makes it less attractive to use gaps occurring in the opposite pedestrian stream.

2. The flow at bottlenecks can be improved by a funnel-shaped construction (see Fig. 6.35(b)) which, at the same time, allows expensive space to be saved. Interestingly, the optimal form resulting from an evolutionary optimization is convex (see Fig. 6.37).

3. A broader door does not necessarily lead to a proportional increase of pedestrian flow through it. Rather, it may lead to more frequent changes of the walking direction which are connected with temporary deadlock situations. Therefore, two doors are much more efficient than one single door with double width. By self-organization, each door is used by one walking direction (cf. Fig. 6.38).

4. Oscillatory changes of the walking direction and periods of standstill in between also occur when different flows *cross* each other. The loss of efficiency caused by this can be reduced by psychological guiding measures or railings initializing roundabout traffic (see Fig. 6.35(c)). Roundabout traffic can already be induced and stabilized by planting a tree in the middle of a crossing, because it suppresses the phases of 'vertical' or 'horizontal' motion in the intersection area. In our simulations this increased efficiency by up to 13%.

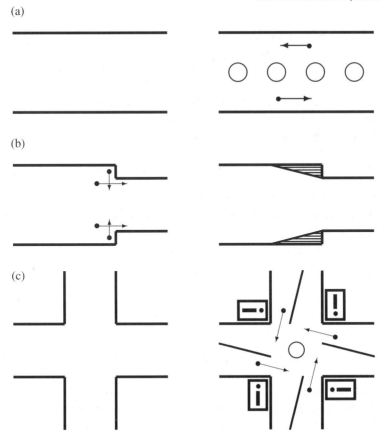

Fig. 6.35 Conventional (left) and improved (right) elements of pedestrian facilities: (a) ways, (b) bottlenecks, and (c) intersections (from [44]). The exclamation marks stand for attraction effects (e.g., interesting posters above the street). Empty circles represent columns or trees, while full circles with arrows symbolize pedestrians and their walking directions.

The complex interaction between various flows can lead to completely unexpected results due to the nonlinearity of the dynamics. (A very impressive and surprising result of evolutionary form optimization is presented in [63].) That means that the planning of pedestrian facilities by conventional methods does not always guarantee the avoidance of big jams, serious obstructions and catastrophic blockages (especially in emergency situations). In contrast, a skillful flow optimization not only enhances efficiency but also saves space that can be used for kiosks, benches or other purposes [44].

6.2.3 Trail formation

Another interesting collective effect of pedestrian motion, which has been investigated very recently, is the development of trail systems on deformable ground. A theory

Fig. 6.36 This photograph of a pedestrian tunnel connecting two subways in Budapest at Deák tér illustrates how a series of columns acts in a similar way to a wall and stabilizes lanes (by preventing their expansion over more than half of the total width of the walkway).

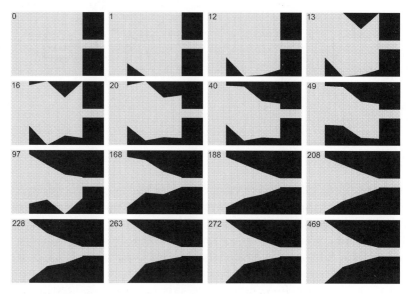

Fig. 6.37 Different phases in the evolutionary optimization of a bottleneck (from [62]).

of human trail formation in green areas like public parks must be able to answer the following questions: Why do pedestrians sometimes build trails in order to save three to five metres, but in other cases accept detours which are much larger? (Fig. 6.39) How and by which mechanism do trail systems evolve in space and time? Why do

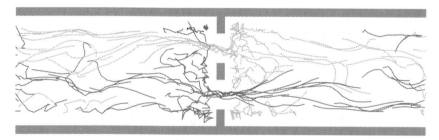

Fig. 6.38 If two alternative passageways are available, pedestrians of opposite walking direction use different doors as a result of self-organization (from [44, 47, 49]).

trails reappear at the same places, if they were destroyed? How should urban planners design public-way systems so that walkers actually use them? Empirical investigations of these questions have, for example, been carried out by Klaus Humpert [64] and Eda Schaur [65].

It is known that many way systems and even streets originated from human or animal trails. For reasons of easier orientation, these mostly point to optically significant places. However, the pedestrians' tendency to take the shortest way to these places and the specific properties of the terrain are often insufficient to explain the trail characteristics. Although trails serve as shortcuts (Fig. 6.40), they frequently do not supply the shortest way to the respective destination. Rather, they are compromises between ways which point in different directions (Fig. 6.41). For example, one often observes a splitting of trails, just before they meet another way in a more or less perpendicular direction (Fig. 6.42). It seems to be much more natural, if trails would form a triangular direct-way system between three entry points and destinations. However, the minimum angle between splitting ways is found to be about 30°.

The active walker model

To simulate the typical features of trail systems, the afore-mentioned behavioural-force model has been extended to an active-walker model together with Joachim Keltsch [8, 47, 66, 67]. Like random walkers, active walkers are subject to the fluctuations and influences of their environment. However, they are additionally able to locally change their environment, e.g., by altering an environmental potential, which in turn influences their further movement and their behaviour. In particular, changes produced by some walkers can influence other walkers. Hence, this nonlinear feedback can be interpreted as an indirect interaction between the active walkers via environmental changes, which may lead to the self-organization of large-scale spatial structures.

Active-walker models have proved their versatility in a variety of applications, such as the formation of complex structures [68], pattern formation in physico-chemical systems [69, 70], aggregation in biological [71] or urban [72] systems, and generation

Fig. 6.39 Sometimes, pedestrians reject detours of only a few metres (above, right hand side), while in other cases they accept much longer detours (below) (from [67]). Obviously, it is the relative, not the absolute, detour that counts. Up to 25% detour seems to be acceptable to pedestrians. If this limit is exceeded, they tend to leave existing ways.

of directed motion [35]. This approach provides a quite stable and fast numerical algorithm for simulating processes involving large density gradients, and it is also applicable in cases where only small particle numbers govern structure formation. In particular, the active walker model is applicable to processes of pattern formation which are intrinsically dependent on the history of their creation.

Fig. 6.40 Pedestrians build trails as shortcuts, if the provided way system is misconceived (from [67]).

First, let us specify how pedestrian motion changes the environment. For this, we represent the ground structure at position r and time t by a function $G(r, t)$ which reflects the comfort of walking. Trails are characterized by particularly large values of G. On the one hand, at their positions $r = r_\alpha(t)$ all pedestrians α leave footprints on the ground (e.g., by trampling down some vegetation). Their intensity is assumed to be $I(r)[1 - G(r, t)/G_{\max}(r)]$, since the clarity of a trail is limited to a maximum value $G_{\max}(r)$. This causes a saturation effect $[1 - G(r, t)/G_{\max}(r)]$ of the ground's alteration by new footprints. On the other hand, the ground structure changes due to the vegetation's regeneration. This will lead to a restoration of the natural ground conditions $G_0(r)$ with a certain weathering rate $1/T(r)$ which is related to the durability $T(r)$ of the trails. Thus, the equation of environmental changes is

$$\frac{dG(r, t)}{dt} = \frac{1}{T(r)}[G_0(r) - G(r, t)] + I(r)\left[1 - \frac{G(r, t)}{G_{\max}(r)}\right]\sum_\alpha \delta(r - r_\alpha(t)), \quad (6.120)$$

where $\delta(r - r_\alpha)$ denotes the Dirac delta function (which yields a contribution only for $r = r_\alpha$).

Now, we will specify how the ground structure influences pedestrian motion. A trail segment at position r motivates a pedestrian at position r_α to move in the direction $(r - r_\alpha)/\|r - r_\alpha\|$. However, its attractiveness decreases with the distance $\|r - r_\alpha(t)\|$. This has been considered by a factor

$$\frac{e^{-\|r - r_\alpha\|/\sigma(r_\alpha)}}{\int d^2 r \, e^{-\|r - r_\alpha\|/\sigma(r_\alpha)}} = \frac{e^{-\|r - r_\alpha\|/\sigma(r_\alpha)}}{2\pi[\sigma(r_\alpha)]^2}, \quad (6.121)$$

Fig. 6.41 These examples of human trails (above, from [44]) and animal trails (below) show that trails are compromises (from [67]). They often deviate from the direct way when they meet other trails.

where $\sigma(\boldsymbol{r}_\alpha)$ characterizes the visibility (so that $1/\sigma$ corresponds to the roughness of the ground). The overall attractive effect of the available trails is obtained by integration of the resulting function over the green area:

$$f_{\mathrm{tr}}(\boldsymbol{r}_\alpha, t) = \int \mathrm{d}^2 r \, \frac{e^{-\|\boldsymbol{r}-\boldsymbol{r}_\alpha\|/\sigma(\boldsymbol{r}_\alpha)}}{2\pi[\sigma(\boldsymbol{r}_\alpha)]^2} G(\boldsymbol{r}, t) \frac{\boldsymbol{r} - \boldsymbol{r}_\alpha}{\|\boldsymbol{r} - \boldsymbol{r}_\alpha\|}. \qquad (6.122)$$

Fig. 6.42 The splitting of a trail which meets another way (and thereby gives rise to an island in the middle) is a rather typical phenomenon (from [67]).

This allows us to define the trail potential

$$V_{tr}(r, t) = - \int_{r_0}^{r} dr' \cdot f_{tr}(r', t) \qquad (6.123)$$

(with an arbitrary but fixed place r_0). Then, the following gradient relation holds:

$$f_{tr}(r, t) = -\nabla V_{tr}(r, t). \qquad (6.124)$$

On homogeneous ground, the walking direction $e_\alpha(r_\alpha, t)$ of a pedestrian α at position $r_\alpha(t)$ is given by the direction of the next destination d_α, i.e.,

$$e_\alpha^0(r_\alpha) = \frac{d_\alpha - r_\alpha}{||d_\alpha - r_\alpha||}. \tag{6.125}$$

However, since the choice of the walking direction e_α is influenced by the destination and existing trails at the same time, we assumed the orientation relation

$$e_\alpha(r_\alpha, t) = \frac{e_\alpha^0(r_\alpha) + f_{tr}(r_\alpha, t)}{||e_\alpha^0(r_\alpha) + f_{tr}(r_\alpha, t)||}. \tag{6.126}$$

This reflects the pedestrians' compromising behaviour, since it specifies the walking direction (which appears in the relaxation term (6.110) of motion) by the average of the direction e_α^0 to the destination and the most attractive walking direction $e_{tr} = f_{tr}/||f_{tr}||$. In cases of rare interactions, the motion of a pedestrian α with desired velocity $v_\alpha^0 \approx v^0$ is simply given by

$$\frac{dr_\alpha}{dt} = v^0 e_\alpha(r_\alpha, t). \tag{6.127}$$

By appropriate scaling, the above equations can be transformed to dimensionless equations which include only two independent parameters $\kappa = IT/\sigma^2$ and $\lambda = v^0 T/\sigma$. Nevertheless, the active-walker model of human trail formation compares very well with empirical findings. This is, for example, illustrated by Fig. 6.43, and by comparing the triangular island in Fig. 6.44(b) with the one in Fig. 6.41(a).

Our simulations begin with spatially homogeneous ground. Pedestrians move between given entry points and destinations at a certain rate, starting at a random point in time. In Fig. 6.43, the entry points and destinations are distributed over the small ends of the ground, while in Fig. 6.44 (Fig. 6.45) pedestrians move between all possible pairs of three (four) fixed places.

At first, pedestrians take the direct ways to their respective destinations, because there is no reason to choose another route (Fig. 6.45 (left)). However, after some time they begin to use already existing paths, since this is more comfortable than to clear new ways. In this way, a kind of selection process [69] between trails sets in. On the one hand, frequently used trails are more comfortable and, therefore, more attractive than others. For this reason they are chosen very often. The resulting reinforcement makes them even more attractive, until the saturation effect becomes effective (due to the limitation of the walking comfort to G_{max}). On the other hand, rarely used trails are destroyed by the weathering effect. This limits the maximum length of the way system which can be supported by a certain rate of trail usage. As a consequence, the trails begin to bundle, especially where different paths meet or intersect. Finally, pedestrians with different destinations use and support common parts of the trail system, which explains the empirically found deviations from direct way systems (Figs. 6.44 (middle) and 6.45 (middle)).

Fig. 6.43 When pedestrians leave footprints on the ground, trails will develop, and only parts of the ground are used for walking (in contrast to paved areas). The similarity between the simulation result (left) and the trail system on the university campus of Brasilia (right) is obvious (photograph by Klaus Humpert) (from [8, 44, 47, 67]).

Fig. 6.44 The structure of the emerging trail system essentially depends on the persistence parameter κ (from [8, 44, 66, 67]). If κ is large, a direct way system results (left). If κ is small, a minimal way system is formed (right). Otherwise a compromise between both extremes will develop (middle), which looks similar to the main trail system in the centre of Fig. 6.41(a).

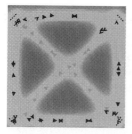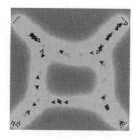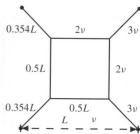

Fig. 6.45 The left and middle graphics illustrate the trail potential $V_{tr}(r, t)$ (from [8, 44, 47, 66, 67]). Arrows represent the positions and directions of pedestrians. These use, with a frequency ν, each of the six connections between the four entry points and destinations in the corners of a square with edge length L. Starting with spatially homogeneous ground, the chosen ways change considerably in the course of time. In the beginning, pedestrians take the direct ways (left). Since frequently used trails become more comfortable, a bundling of trails sets in which reduces the overall length of the trail system (middle). The finally resulting way system could serve as a planning guideline (right). It provides a suitable compromise between small construction costs and large comfort of walking. At an overall length which is 50% shorter than the direct way system, it requires everyone to take a relative detour of 21%, which is a fair and acceptable solution. Moreover, the resulting trail system is structurally stable under the assumption that a frequency of usage of 1.5ν is needed to support permanent trails in competition with the weathering effect. If the frequency of usage is higher than this threshold, the outcome is expected to be a direct way system.

A direct-way system (which provides the shortest connections, but covers a lot of space) only develops in cases of high frequencies of usage and almost equal walking comfort for all routes (Fig. 6.44 (left)). If the persistence κ of existing trails is small (e.g., for a rapidly regenerating ground), the final trail system is a minimal way system (which is the shortest-way system that connects all entry points and destinations) (Fig. 6.44 (right)). For realistic values of κ, the evolution of the trail system stops before this state is reached (Fig. 6.44 (middle)). Thus, κ is related to the average relative detour of the walkers.

Optimization of way systems

The trail systems resulting from the above described mechanisms are particularly suited as planning guidelines for the construction of optimal way systems. First of all, they take into account the walking and orientation habits, so that pedestrians will actually use such ways. Second, the resulting trail systems provide the best compromise between maximum shortness and comfort of ways. Third, they seem to offer fair solutions, which balance the relative detours of all pedestrians (Fig. 6.45 (right)).

Computer simulations of this kind can be used for answering various questions, given a knowledge of the entry points and destinations as well as the expected rates of usage of the corresponding connections (which can be determined by established models [41]):

- Which is the trail system that pedestrians would naturally use? *Solution:* Simulate the problem with realistically chosen values of λ and κ.

- Is the resulting way system structurally stable with respect to small changes of the parameters λ and κ (e.g., due to varying weather conditions)? *Solution:* Simulate the system with slightly modified parameter values and see whether the topological structure of the trail system changes.

- Given a certain amount of money which allows a way system of a certain length to be, which way system should be built, i.e., which one is most comfortable or 'intelligent'? *Solution:* Control the overall length of the evolving way system by varying κ until it fits the desired length.

- If a certain level of comfort shall be provided, which is the cheapest way system fulfilling this demand? *Solution:* Increase κ, starting with small values, until the pedestrians take the average relative detour which was specified to be acceptable.

- Given an existing way system, how should it be extended? *Solution:* Take into account the existing way system by setting $G_0(r) = G_{max}$ and see where the resulting way system contains additional trails.

Due to the large number of walkers and the different time scales involved, the above proposed multi-agent simulations of trail formation are very time consuming. Therefore, it is reasonable to derive macroscopic equations of trail formation from the microscopic ones which were discussed, here. Since the finally evolving trail systems correspond to the stationary solutions of these macroscopic equations, they can be obtained by a simple and fast iteration scheme. A detailed discussion of this issue is presented in [35].

Trail formation by animals

Trails are not only formed by pedestrians, but also by hoofed animals (Figs. 6.41(b) and 6.46). Even mice leave trail systems on the ground [44,65] (Fig. 6.47). A similar thing is observed for ants, which build trunk-like trails [44,74] (Fig. 6.48). However, ant trails are based on chemotaxis, i.e., on chemical markings instead of footprints. A simulation model for trunk trail formation on the basis of two different kinds of chemicals has been proposed by Frank Schweitzer *et al.* [35].

6.2.4 Conclusions

It was pointed out that pedestrian dynamics shows various collective phenomena. These can be interpreted as self-organization effects due to nonlinear interactions among pedestrians. The empirical findings can be realistically described by molecular-dynamic simulations of pedestrian streams which are based on a behavioural-force model. Some of the phenomena, like lane formation or oscillatory flows through bottlenecks, show striking similarities with granular flows. Applications to the optimization of pedestrian facilities are quite natural. A clever use of self-organized patterns of motion even allows us to achieve more efficient pedestrian

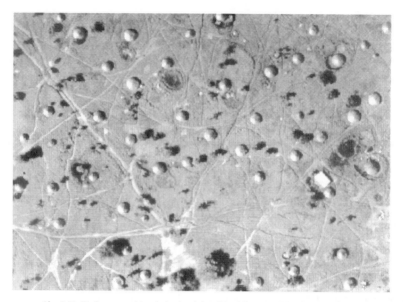

Fig. 6.46 Trail system of hoofed animals in Africa (photograph by Georg Gerster).

Fig. 6.47 Part of the trail system of mice after photographs made in Warmbronn, Germany (reproduction with kind permission of Frei Otto [73]).

flows with less space. In addition, improvements of way systems can be worked out with an active-walker model of human trail formation. This includes additional indirect interactions between pedestrians which are caused by environmental changes and their influence on human walking behaviour.

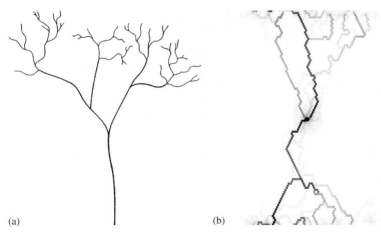

Fig. 6.48 (a) Dendritic trunk trail system of the ant species *Pheidole militicida* (after [74]). (b) Simulation result of trunk trail formation by active walkers. The result is in good agreement with the empirical findings (from [35].)

Acknowledgements

D.H. wishes to thank the DFG for financial support (Heisenberg scholarship He 2789/1-1). Moreover, the authors are grateful to Kai Bolay, Illés Farkas, Georg Gerster, Klaus Humpert, Joachim Keltsch, Frei Otto, Frank Schweitzer, and Tamás Vicsek for their fruitful collaboration and/or for providing various pictures.

References

[1] C. W. Reynolds. Flocks, herds, and schools: a distributed behavioral model. *Comp. Graph.* 21: 25–34, 1987.

[2] J. K. Perrish and L. Edelstein-Keshet. Complexity, pattern and evolutionary trade-offs in animal aggregation. *Science* 284: 99–101, 1999.

[3] E. M. Rauch, M. M. Millonas, and D. R. Chialvo. Pattern formation and functionality in swarm models. *Phys. Lett. A* 207: 185, 1995.

[4] H. E. Stanley. *Introduction to phase transitions and critical phenomena*. Oxford University Press, Oxford, 1971.

[5] S.-K. Ma. *Statistical mechanics*. World Scientific, Singapore, 1985.

[6] T. Vicsek, A. Czirók, E. Ben-Jacob, I. Cohen, and O. Shochet. Novel type of phase transition in a system of self-driven particles. *Phys. Rev. Lett.* 75: 1226–1229, 1995.

[7] N. Shimoyama, K. Sugawara, T. Mizuguchi, Y. Hayakawa, and M. Sano. Collective motion in a system of motile elements. *Phys. Rev. Lett.* 76: 3870–3873, 1996.

[8] D. Helbing, J. Keltsch, and P. Molnar. Modelling the evolution of human trail systems. *Nature* 388: 47–50, 1997.

[9] N. D. Mermin and H. Wagner. Absence of ferromagnetism or antiferromagnetism in one- or two-dimensional isotropic Heisenberg models. *Phys. Rev. Lett.* 17: 1133, 1966.

[10] A. Czirók, H. E. Stanley, and T. Vicsek. Spontaneously ordered motion of self-propelled particles. *J. Phys. A* 30: 1375–1385, 1997.

[11] J. Toner and Y. Tu. Long-range order in a two-dimensional dynamical xy model: how birds fly together. *Phys. Rev. Lett.* 75: 4326–4329, 1995.

[12] R. B. Stinchcombe. *Phase Transitions and Critical Phenomena*, Vol. 7. Academic Press, New York, 1983.

[13] J. M. Kosterlitz and D. J. Thouless. Ordering, metastability and phase transitions in two-dimensional systems. *J. Phys. C.* 6: 1181, 1973.

[14] A. Czirók, M. Vicsek, and T. Vicsek. Collective motion of organisms in three dimensions. *Physica A* 264: 299–304, 1999.

[15] A. Czirók, A.-L. Barabási, and T. Vicsek. Collective motion of self-propelled particles: kinetic phase transition in one dimension. *Phys. Rev. Lett.* 82: 209–212, 1999.

[16] A. Czirók, E. Ben-Jacob, I. Cohen, and T. Vicsek. Formation of complex bacterial patterns via self-generated vortices. *Phys. Rev. E* 54: 1791–1801, 1996.

[17] Y. L. Duparcmeur, H. J. Herrmann, and J. P. Troadec. Spontaneous formation of vortex in a system of self-motorized particles. *J. Phys. (France) I* 5: 1119–1128, 1995.

[18] J. Hemmingsson. Modellization of self-propelling particles with a coupled map lattice model. *J. Phys. A* 28: 4245–4250, 1995.

[19] Z. Csahók and T. Vicsek. Lattice-gas model for collective biological motion. *Phys. Rev. E* 52: 5297, 1995.

[20] R. B. Potts. *Proc. Camb. Phil. Soc.* 48:106, 1952.

[21] K. Binder, K. Vollmayr, H.-P. Deutsch, J. D. Reger, and M. Sheucher. *Int. J. Mod. Phys. C* 5: 1025, 1992.

[22] E. V. Albano. Self-organized collective displacements of self-driven individuals. *Phys. Rev. Lett.* 77: 2129 – 2132, 1996.

[23] F. Y. Wu. The Potts model. *Rev. Mod. Phys.* 54: 235, 1992.

[24] K. Binder. Static and dynamic critical phenomena of the two-dimensional q-state Potts model. *J. Stat. Phys.* 24: 69, 1981.

[25] H. J. Bussemaker, A. Deutsch, and E. Geigant. Mean-field analysis of a dynamical phase transition in a cellular automaton model for collective motion. *Phys. Rev. Lett.* 78: 5018–5021, 1997.

[26] H. J. Bussemaker. Analysis of a pattern-forming lattice-gas automaton: mean-field theory and beyond. *Phys. Rev. E* 53: 1644, 1996.

[27] A. Czirók. *Models of collective behaviour in biology.* (PhD thesis, Eötvös University, Budapest, 2000).

[28] Z. Csahók and A. Czirók. Hydrodynamics of bacterial motion. *Physica A* 243: 304, 1997.

[29] M. Abramowitz and I. A. Stegun (ed.). *Handbook of Mathematical Functions.* Dover Publications, New York, 1970.

[30] J. Toner and Y. Tu. Flocks, herds, and schools: a quantitative theory of flocking. *Phys. Rev. E* 58: 4828, 1998.

[31] D. Forster, D. R. Nelson, and M. J. Stephen. Large-distance and long-time properties of a randomly stirred fluid. *Phys. Rev. A* 16: 732, 1977.

[32] K. L. Schick and A. A. Verveen. $1/f$ noise with a low frequency white noise limit. *Nature* 251: 599, 1974

J. A. C. Gallas, H. J. Herrmann and S. Sokołowski. Convection cells in vibrating granular media. *Phys. Rev. Lett.* 69: 1371-1374, 1992.

H. A. Makse, S. Havlin, P. R. King and H. E. Stanley. Spontaneous stratification in granular mixtures. *Nature* 386: 379–382, 1997.

[33] H. J. Herrmann. In: P.L. Garrido, J. Marro (ed.) *3rd Granada Lectures in Computational Physics.* (Springer, Heidelberg, 1995).

[34] E. Ben-Jacob, O. Shochet, A. Tenenbaum, I. Cohen, A. Czirók and T. Vicsek. Generic modelling of cooperative growth patterns in bacterial colonies. *Nature* 368: 46–49, 1994.

D.A. Kessler, H. Levine. Pattern formation in Dictyostelium via the dynamics of cooperative biological entities. *Phys. Rev. E* 48: 4801–4804, 1993.

[35] D. Helbing, F. Schweitzer, J. Keltsch, P. Molnár. Active-walker model for the formation of human and animal trail systems. *Phys. Rev. E* 56: 2527–2539, 1997.

[36] W. Weidlich and G. Haag. *Concepts and Models of a Quantitative Sociology.* Springer, Berlin, 1983.

W. Weidlich. Physics and social science – the approach of synergetics. *Phys. Rep.* 204: 1, 1991.

D. Helbing. *Quantitative Sociodynamics. Stochastic Methods and Models of Social Interaction Processes.* Kluwer Academic, Dordrecht, 1995.

[37] *Highway Capacity Manual,* Chap. 13. Special Report 209. Transportation Research Board, Washington, D.C., 1985.

[38] U. Weidmann. *Transporttechnik der Fußgänger.* Schriftenreihe des Instituts für Verkehrsplanung, Transporttechnik, Straßen- und Eisenbahnbau Nr. 90, ETH Zürich, 1993.

[39] D. Oeding. *Verkehrsbelastung und Dimensionierung von Gehwegen und anderen Anlagen des Fußgängerverkehrs.* Straßenbau und Straßenverkehrstechnik, Heft 22, Bonn, 1963.

S. J. Older. Movement of pedestrians on footways in shopping streets. *Traffic Eng. Control* 10: 160, 1968.

P. D. Navin and R. J. Wheeler. Pedestrian-flow characteristics. *Traffic Eng.* 39: 31, 1969.

[40] A. J. Mayne. Some further results in the theory of pedestrians and road traffic. *Biometrica* 41: 375, 1954.

N. Ashford, M. O'Leary, P. D. McGinity. Stochastic modelling of passanger and baggage flows through an airport terminal. *Traffic Eng. Control* 17: 207, 1976.

S. J. Yuhaski Jr. and J. M. Smith. Modelling circulation systems in buildings using state-dependent queueing models. Queueing Systems 4: 319, 1989.

G. G. Løvås. In A. Pave (ed.) *Modelling and Simulation 1993*. Society for Computer Simulation International, Ghent, Belgium, 1993.

[41] A. Borgers, H. J. P. Timmermans. City centre entry points, store location patterns and pedestrian route choice behaviour: a microlevel simulation model. *Socio-Economic Planning Science* 20: 25, 1986.

H. Timmermans, X. van der Hagen, and A. Borgers. Transportation systems, retail enviroments and pedestrian trip chaining behaviour: modelling issues and applications. *Transportation Res. B* 26: 45–59, 1992.

[42] L. F. Henderson. On the fluid mechanics of human crowd motion. *Transportation Res.* 8: 509, 1974.

[43] D. Helbing. A fluid-dynamic model for the movement of pedestrians. *Complex Systems* 6: 391, 1992.

D. Helbing. *Stochastische Methoden, nichtlineare Dynamik und quantitative Modelle sozialer Prozesse*. Shaker, Aachen, Germany, 1993.

[44] D. Helbing. *Verkehrsdynamik. Neue physikalische Modellierungskonzepte*. Springer, Berlin, 1997.

[45] D. Helbing. A mathematical model for the behavior of pedestrians. *Behavioral Science* 36: 298, 1991.

[46] D. Helbing, P. Molnár. Social force model for pedestrian dynamics. *Phys. Rev. E* 51: 4282, 1995.

[47] P. Molnár. *Modellierung und Simulation der Dynamik von Fußgängerströmen* (Shaker, Aachen, Germany, 1996).

[48] D. Helbing. In D. E. Wolf, M. Schreckenberg and A. Bachen (ed.) *Traffic and Granular Flow*. World Scientific, Singapore, 1996.

[49] D. Helbing, P. Molnár. In F. Schweitzer (ed.) *Self-Organization of Complex Structures: From Individual to Collective Dynamics*. Gordon and Breach, London, 1997.

[50] P. G. Gipps and B. Marksjö. A micro-simulation model for pedestrian flows. *Mathematics and Computers in Simulation* 27: 95, 1985.

[51] Video films by T. Arns. Wannenstr. 22, 70199 Stuttgart, Germany.

[52] L. F. Henderson. The statistics of crowd fluids. *Nature* 229: 381, 1971.

L. F. Henderson and D. J. Lyons. Sexual differences in human crowd motion. *Nature* 240: 353, 1972.

L. F. Henderson and D. M. Jenkins. Response of pedestrians to traffic challenge. *Transportation Res.* 8: 71, 1973.

[53] J. S. Coleman, and J. James. The equilibrium size distribution of freely-forming groups. *Sociometry* 24: 36, 1961.

J. S. Coleman. *Introduction to Mathematical Sociology*. The Free Press of Glencoe, New York, 1964.

[54] S. B. Santra, S. Schwarzer, and H. Herrmann. Fluid-induced particle-size segregation in sheared granular assemblies. *Phys. Rev. E* 54: 5066, 1996.

[55] K. Yoshikawa, N. Oyama, M. Shoji, and S. Nakata. Use of a saline oscillator as a simple nonlinear dynamical system: rhythms, bifurcation, and entrainment. *Am. J. Phys.* 59: 137, 1991.

X.-L. Wu, K. J. Måløy, A. Hansen, M. Ammi, and D. Bideau. Why hour glasses tick. *Phys. Rev. Lett.* 71: 1363, 1993.

T. L. Pennec, K. J. Måløy, A. Hansen, M. Ammi, D. Bideau, and X.-L. Wu. Ticking hour glasses: experimental analysis of intermittent flow. *Phys. Rev. E* 53: 2257, 1996.

[56] K. Lewin. *Field Theory in Social Science.* Harper and Brothers, New York, 1951.

[57] D. Helbing. Boltzmann-like and Boltzmann–Fokker–Planck equations as a foundation of behavioral models. *Physica A* 196: 546, 1993; and *J. Math. Sociology* 19: 189, 1994.

[58] D. Helbing, and T. Vicsek. Optimal self-organization. *New J. Phys.* 1: 13.1–13.17, 1999.

[59] D. Helbing, I. Farkas, and T. Vicsek. Freezing by heating in a driven mesoscopic system. *Phys. Rev. Lett.* 84: 1240–1243, 2000.

[60] D. Helbing. In *Natural Structures,* Part II. Sonderforschungsbereich 230, Stuttgart, Germany, 1992, p. 93.

[61] I. Rechenberg. *Evolutionsstrategie: Optimierung technischer Systeme nach Prinzipien der biologischen Evolution.* Frommann-Holzboog, Stuttgart, 1973.

[62] K. Bolay. *Nichtlineare Phänomene in einem fluid-dynamischen Verkehrsmodell* (Master's thesis, University of Stuttgart, Germany, 1999).

[63] J. Klockgether, H. P. Schwefel. In *Proc. 11. Symp. on Eng. Aspects of MHD.* Cal. Inst. Techn., 1970, p. 141.

[64] K. Humpert *et al.* In K. Teichmann and J. Wilke (ed.). *Prozeß und Form 'Natürlicher Konstruktionen'. Der Sonderforschungsbereich 230.* Ernst & Sohn, Berlin, 1996, p. 172.

[65] E. Schaur. *Ungeplante Siedlungen / Non-Planned Settlements.* Krämer, Stuttgart, Germany, 1991.

[66] J. Keltsch. *Selbstorganisation von Wegen durch 'Active Walkers'* Master's thesis, University of Stuttgart, Germany, 1996.

[67] D. Helbing. In M. Schreckenberg and D. Wolf (ed.), *Traffic and Granular Flow '97.* Springer, Singapore, 1998.

[68] B. Davis. Reinforced Random Walk. *Probab. Th. Rel. Fields* 84: 203, 1990.

R. D. Freimuth and L. Lam, In L. Lam and V. Naroditsky (ed.), *Modeling Complex Phenomena.* Springer, New York, 1992.

D. R. Kayser, L. K. Aberle, R. D. Pochy and L. Lam. Active-walker models: tracks and landscapes. *Physica A* 191: 17, 1992.

R. D. Pochy, D. R. Kayser, L. K. Aberle and L. Lam. Boltzmann active walkers and rough surfaces. *Physica D* 66: 166, 1993.

L. Lam, and R. Pochy. Active-walker models: growth and form in nonequilibrium systems. *Computers in Physics* 7: 534, 1993.

L. Lam. Active-walker models for complex systems. *Chaos, Solitons & Fractals* 6: 267, 1995.

F. Schweitzer. In L. Schimansky-Geier and T. Pöschel (ed.). *Lectures on Stochastic Dynamics*. Springer, Berlin, 1997.

[69] F. Schweitzer and L. Schimansky-Geier. Clustering of 'active' walkers in a two-component system. *Physica A* 206: 359, 1994.

[70] L. Schimansky-Geier, M. Mieth, H. Rosé and H. Malchow. Surface formation by active Brownian particles. *Phys. Lett. A* 207: 140, 1995.

L. Schimansky-Geier, F. Schweitzer and M. Mieth. In F. Schweitzer (ed.), *Self-Organization of Complex Structures: From Individual to Collective Dynamics*. Gordon and Breach, London, 1997, p. 101.

[71] A. Stevens. *Mathematical Modeling and Simulations of the Aggregation of Myxobacteria*. PhD thesis, Ruprecht-Karls-University Heidelberg, 1992.

A. Stevens and F. Schweitzer. In W. Alt, A. Deutsch and G. Dunn (ed.), *Dynamics of Cell and Tissue Motion*. Birkhäuser, Basel, 1997, p. 183.

[72] F. Schweitzer and J. Steinbrink. In F. Schweitzer (ed.), *Self-Organization of Complex Structures: From Individual to Collective Dynamics*. Gordon and Breach, London, 1997, p. 501.

[73] F. Otto. *Die natürliche Konstruktion gewachsener Siedlungen*. Sonderforschungsbereich 230, Stuttgart, Germany, 1991, Heft 37.

[74] B. Hölldobler and E. O. Wilson. *The Ants*. Belknap, Cambridge, MA, 1990.

Index